The City Creative

The Rise of
Urban Placemaking
in Contemporary
America

The City Creative

MICHAEL H. CARRIERE &
DAVID SCHALLIOL

The University of Chicago Press   Chicago and London

The University of Chicago Press, Chicago 60637
The University of Chicago Press, Ltd., London
© 2021 by The University of Chicago
All rights reserved. No part of this book may be used or reproduced in any manner
whatsoever without written permission, except in the case of brief quotations
in critical articles and reviews. For more information, contact the University of
Chicago Press, 1427 E. 60th St., Chicago, IL 60637.
Published 2021
Printed in China

30 29 28 27 26 25 24 23 22 21   1 2 3 4 5

ISBN-13: 978-0-226-72722-6 (cloth)
ISBN-13: 978-0-226-72736-3 (e-book)
DOI: https://doi.org/10.7208/chicago/9780226727363.001.0001

Library of Congress Cataloging-in-Publication Data

Names: Carriere, Michael H., author. | Schalliol, David, author.
Title: The city creative : the rise of urban placemaking in contemporary America /
    Michael H. Carriere and David Schalliol.
Description: Chicago ; London : The University of Chicago Press, 2021. | Includes
    bibliographical references and index.
Identifiers: LCCN 2020013473 | ISBN 9780226727226 (cloth) | ISBN
    9780226727363 (ebook)
Subjects: LCSH: City planning—United States—History—21st century. | City
    planning—United States—History—20th century. | Community development,
    Urban—United States. | City planning—Citizen participation. | Urban renewal—
    Citizen participation. | Public spaces—Social aspects—United States.
Classification: LCC HT167 .C37 2021 | DDC 307.1/21609730905—dc23
LC record available at https://lccn.loc.gov/2020013473

♾ This paper meets the requirements of ANSI/NISO Z39.48-1992
(Permanence of Paper).

Contents

A Brief History of the Recent Past

(*Previous spread*) Detroit's Belle Isle on the day the city declared bankruptcy, July 18, 2013.

If I could do it, I'd do no writing at all here. It would be photographs; the rest would be fragments of cloth, bits of cotton, lumps of earth, records of speech, pieces of wood and iron, phials of odors, plates of food and of excrement.

James Agee (1941)

In August 2010, Quicken Loans founder Dan Gilbert moved the headquarters of his company—and its seventeen hundred employees—from the suburbs of Detroit to the city's troubled downtown. At the time, the move stunned many observers as Detroit was still reeling from the effects of the Great Recession and the bankruptcies of Chrysler and General Motors. At the same time, Gilbert's Bedrock Real Estate Services went on a buying spree, gobbling up countless historic buildings, empty skyscrapers, and myriad other properties throughout the city. In December 2017, *Crain's Detroit Business* forecasted that Bedrock would have over twenty-four million square feet under development by 2022, at an estimated value of approximately $4 billion.[1]

This level of investment in an American urban center was noteworthy in and of itself. Yet what was even more stunning was the way Gilbert began to use such investment as a means of redefining what it meant to live and work in downtown Detroit. The real estate, in other words, was just the first step in a broader process of reinventing urban living in a city still carving out its place in the twenty-first century. Still suffering from the aftershocks of deindustrialization, white flight, and more, Detroit's rebranding would not come easy, particularly when—as would come back to haunt Gilbert—many of the neighborhoods hit hardest would have to wait.

But Gilbert had help. By early 2013, he was in close contact with Fred Kent, who in 1975 founded the New York–based Project for Public Spaces (PPS), a nonprofit organization that declares itself to be "dedicated to helping create and sustain public spaces that build strong communities." The early twenty-first century found PPS employing the concept of creative placemaking as perhaps the best mechanism with which to rebuild America's urban centers. For PPS, placemaking was "a collaborative process by which we can shape our public realm in order to maximize shared value." While paying attention to the traditional world of city planning, placemaking is "more than just promoting better urban design." Instead, it "facilitates creative patterns of use, paying particular attention to the physical, cultural, and social identities that define a place and support its ongoing evolution." And Detroit proved an appealing laboratory for such ideas, with Gilbert a more-than-willing partner. In fact, he would speak at PPS's inaugural Placemaking Leadership Council, held in Detroit in April 2013. Not surprisingly, his topic was the ability of placemaking to transform his city.[2]

Kent would work with Gilbert and others on "Opportunity Detroit: A Placemaking Vision for Downtown Detroit," a report issued in the spring of 2013. This report called for a "culture change" in city planning, one based on the core belief that an "intense focus on the public realm will transform streets, sidewalks, promenades and buildings so that they relate to pedestrians on a human scale." Rather than focusing on economic activity, it stressed the need to grow social activity—all sorts of development, including economic, would soon follow. "The downtown core," it noted, "will become all about activity on the streets, sidewalks, parks and plazas that draw more and more people." The ultimate goal was to create spaces of "sociability." "When people see friends, meet and greet their neighbors, and feel comfortable interacting with strangers," the report concluded, "they tend to feel a stronger sense of place or attachment to their community—and to the place that fosters these types of social activities."[3]

In the aftermath of this report, Gilbert became the city's chief placemaker-in-residence. He placed whimsical furniture in front of buildings and created a bike rental program for his workers. He strung festive lights along Woodward Avenue and gave food trucks and street performers carte blanche to take advantage of underutilized spaces near his properties. He covered projects like his Z Garage with artwork and even hired the acclaimed artist Shepard Fairey to paint a mural on the former Compuware Building. And under the influence of Fred Kent—who had seen a similar project in Paris—he installed a beach on Campus Martius, a park in downtown Detroit that has served as the epicenter of the city's placemaking efforts. These projects are meant to appeal to the thousands of young people now working in downtown Detroit. By 2017, for example, over twelve thousand of Quicken Loans' sixteen thousand total employees worked in the company's downtown office. Such results led Kent to remark about Gilbert's vision for Detroit: "It is the most extraordinary project that I have worked on in my entire career."[4]

While Gilbert casts himself as a man of action, it is clear that his work and the report that has served as his placemaking blueprint drew from a host of intellectual sources. "Opportunity Detroit" cites the work of the sociologist William H. Whyte—an important early influence on the development of PPS—and his praise of "small spaces" in large cities. It quoted Whyte extensively on the value of such spaces, including ideas like this: "It is not just the number of people using them, but the larger number who pass by and enjoy them vicariously, or even the larger number who feel better about the city center for knowledge of them." Such ideas undoubtedly informed Gilbert's approach to projects like Campus Martius.[5]

There is also little doubt that Gilbert leaned heavily on the ideas of the urban theorist Richard Florida and his belief in the transformative power of the *creative class*. Florida felt that cities that could shape their built environment, often through the arts and other cultural tools, put themselves in a better position to attract both the companies and the employees that drove the creative economy—and therefore urban economic growth in the twenty-first century. And he was clearly a fan of Gilbert's work. In May 2012, he cited Gilbert's commitment to "spurring the revitalization of the city's downtown core" as a crucial component of how Detroit was "rising."[6]

In examining the rise of Detroit, Florida would also note the influence of the vaunted author/activist Jane Jacobs on the efforts of Gilbert and his associates. To Florida, Gilbert's repurposing of once-abandoned buildings proved that he was attuned to the advice that Jacobs offered in her foundational 1961 *The Death and Life of Great American Cities*: "New ideas must use old buildings." In 2016, the *Detroit Free Press* found "Jane Jacobs' influence inspiring" projects throughout Detroit, while the blog *Deadline Detroit* enthusiastically reported that Gilbert's "downtown gambit" was destined to succeed "because his placemaking efforts are largely based on the ideas of Jane Jacobs, the greatest urban planning critic ever." Importantly, Gilbert did not shy away from such associations. In the brochure meant to commemorate the launch of City Modern, a mixed-use, pedestrian-friendly community development located in Detroit's Brush Park, Gilbert made clear the influence of Jacobs on the residential project by including a quote from her 1958 essay "Downtown Is for People" on the front of the document: "Designing a dream city is easy; rebuilding a living one takes imagination."[7]

And projects like City Modern were only pieces of a broader reimagining of the city, the creation of what Gilbert referred to as "Detroit 2.0." In explaining what this was, he highlighted how spaces of culture and sociability would lead to what he truly wanted: a bustling new economy for Detroit. In a 2011 blog post, he wrote that he strove to create an urban environment that is "hustling and bustling with young technology-focused people who can find cool lofts to live in, abundant retail and entertainment options close by, safe streets day and night, and most importantly, numerous start-up and growing entrepreneurial companies where opportunity is endless and creative minds are free to collaborate and do what they do best: CREATE!"[8]

Very quickly, and somewhat surprisingly, Gilbert's investment in placemaking begat more investment. Myriad government agencies, grantmaking organizations, philanthropic bodies, and private investors came to town to fund even more placemaking activity. At the national level, ArtPlace America—a consortium led by the National Endowment for the Arts (NEA) and made up of other federal agencies, foundations, and financial institutions—provided close to $5 million in grant money to at least fourteen placemaking projects in Detroit between 2011 and 2017. "Our Town," another NEA grantmaking program specifically geared toward creative-placemaking projects, has offered six-figure grants to various placemakers in Detroit, including a $100,000 grant given to the Detroit Economic Growth Association in 2013. Through these efforts and their ripple effects, there is no doubt that select local actors were benefiting from the money being steered to creative-placemaking work.[9]

The document that fueled the NEA's understanding of creative placemaking in Detroit and elsewhere—a 2010 white paper authored for the agency by the economist Ann Markusen and the arts administrator turned urban planner Anne Gadwa Nicodemus—dovetailed nicely with Gilbert's approach to placemaking. The "Creative Placemaking" paper synthesized and streamlined the movement's intellectual sources into an easy-to-digest document. And, like Gilbert and Florida, Markusen and Gadwa Nicodemus also made the case that sites of urban sociability lead to economic growth. With the NEA offering Markusen and Gadwa Nicodemus a national platform, their report quickly became ground zero for the explosion of creative-placemaking activity across the country.

If nothing else, Markusen and Gadwa Nicodemus's paper offered a clear yet expansive definition: "Creative placemaking animates public and private spaces, rejuvenates structures and streetscapes, improves local business viability and public safety, and brings diverse people together to celebrate, inspire, and be inspired. In turn, these creative locales foster entrepreneurs and cultural industries that generate jobs and income." In contrast to earlier, highly centralized approaches to urban redevelopment, creative placemaking, Markusen and Gadwa Nicodemus asserted, "envisions a more decentralized portfolio of spaces acting as creative crucibles." More specifically,

such sites provide useful amenities for "American cultural industries," thereby helping attract more and more valuable "Creative Workers and Entrepreneurs." Not surprisingly, the white paper moved away from the narrative of cities as centers of industrial activity while paying great attention to sites that privileged the service economy and artistic production. There is little wonder that such a model proved attractive to Gilbert and Detroit's political and economic leadership.[10]

Yet what is perhaps most noteworthy about the rise of creative placemaking is the way that it has become the logic—indeed, the very ethos—of a broad system of urban redevelopment, one that in fact transcends the very physicality of urban spaces. As the Massachusetts Institute of Technology urbanist Susan Silberberg argued in an influential 2013 paper: "The most successful placemaking initiatives transcend the 'place' to forefront the 'making.'" In the process, these efforts privileged "process over product": "In placemaking, the important transformation happens in the minds of the participants, not simply the space itself." Similarly, a November 2014 report from the NEA and ArtPlace America urged creative placemakers to move "beyond the building." According to the report, "creative placemaking does not just refer to physical/built spaces" as placemakers are often striving to "get beyond their physical space."[11]

This diminishment in importance of the physicality of creative placemaking allows practitioners to suggest that the process can be applied anywhere and that it permits the creation of what can best be understood as the intellectual framework for such development work. Creative placemaking can, in the words of Silberberg, "facilitate social interaction," and the process can lead to "third places" that "foster civic connections and build social capital." At the heart of such concepts is a firm belief in the transformative potential of cultural production and the act of consuming such cultural products. The importance of local history and conditions—along with the narrative of producing anything other than art and/or spaces of sociability—is often significantly downplayed.[12]

This one-size-fits-all model, divorced in large part from the messiness of differing local environments, has allowed creative placemaking to take root in urban centers across the United States. Detroit, in other words, is anything but exceptional. Government bodies at the national, state, and local levels alike have turned to creative placemaking as a central redevelopment tool in the early twenty-first century, with some—in cities such as Atlanta, Minneapolis, and Washington, DC, and states such as Connecticut and Indiana—establishing official placemaking offices and programs. And, like Detroit, these actors are taking advantage of funding offered by organizations like the NEA and ArtPlace America. At the philanthropic foundation level alone, the Kinder Foundation, the Kresge Foundation, the Ewing Marion Kauffman Foundation, the William Penn Foundation, and the Educational Foundation of America all have focused on creative placemaking. Private-sector actors like Blue Cross Blue Shield and Southwest Airlines have also embraced creative placemaking. Across the country, conferences and publications on the topic of creative placemaking have been sponsored by PPS (New York), the Urban Land Institute (Washington, DC), the Federal Reserve Bank of San Francisco, and the Institute for Quality Communities (Norman, Oklahoma).[13]

All this activity leaves no doubt that interest in such strategies has exploded over the past decade. Why now? The answer lies in the conditions created by the Great Recession. The realities of the recession hit every major city and disrupted the economic status quo in such places more so than any financial catastrophe since the Great Depression. At the broadest level, the recession brought forward an intense effort to reexamine the workings of capitalism. For those like David Harvey and Mark Blyth, the economic crisis was an extreme—though perhaps logical—result of histories driven by such concepts as neoliberalism and austerity. It was the end of something and the beginning of new understandings of economic life.[14]

It is in this context that we must begin to understand the contemporary rise of creative placemaking. Creative placemaking grew in popularity as certain actors, including individuals, not-for-profit organizations, and municipalities, used the practice to address the intense upheaval ushered in by the recession. In very real ways, then, we see how creative placemaking emerged as a strategy to address the crises—and how the practice can provide a new way to think about political economy. More specifically, we—like Harvey and others—see the city as the place where the damage wrought by economic crisis is most visible. At the same time, it was during this moment that the problems facing the American city became glaringly apparent to more and more city dwellers and activists began to attempt to solve these problems in innovative ways. In the immediate aftermath of the recession, creative placemaking provided the tools for such work.[15]

Other advocates of creative placemaking have voiced similar arguments. As Markusen and Gadwa Nicodemus concluded in their influential 2010 white paper: "In this difficult Great Recession era, creative placemaking has paradoxically quickened." Expounding on this phenomenon in a November 2013 policy brief for the Ewing Marion Kauffman Foundation, Markusen explained: "Since the Great Recession, North American mayors and city councils have boosted investments in arts and culture as creative placemaking to improve the quality of life, to attract residents, managers, and workers, and to welcome visitors." With an eye on attempting to renew postrecession urban economies, she described city leaders "newly aware that artists bring income into the city, improve the performance of area businesses and creative industries, and directly create new businesses and jobs."[16]

Obviously, we are not the first to take notice of the growing popularity of creative placemaking. We are, however, the first to give this rise the history it deserves. The still unfolding consequences of the Great Recession, along with the "newly aware" status of city leaders and other decisionmakers to the perceived economic potential of the arts and culture, has led to a somewhat ahistorical understanding of creative placemaking. It has been cast as a novel approach to solving an unprecedented crisis and hailed as a catalyst for more than a decade of urban economic expansion. Very little attention has been paid to the history behind such efforts, which has produced a lack of critical attention to this growing phenomenon. Of course, there are voices that challenge the mainstream approach to creative placemaking, particularly in social practice art and in the fields of race and gentrification studies, but these critiques are often treated as asides about the importance of marginalized voices in an otherwise traditional narrative. Investors, foundations, and governments—those that came to both fund and practice creative placemaking—have rarely made these ideas central to the identity of placemaking itself. For every placemaker like the city of Oakland's cultural affairs manager, Roberto Bedoya, who has argued for recognizing dispossession as a common part of the placemaking process, there are numerous others proceeding with business as usual.[17] What we see instead are often overly celebratory accounts of the work of placemakers like Dan Gilbert, supplanted by cursory accounts of such influential figures as Jane Jacobs—and all presented with an unblinking faith in the power of individual creativity and the collective ability of the city to foster sociability, which in turn fosters entrepreneurialism and economic development.[18]

This book begins by offering a history behind this contemporary moment, providing a narrative for what can best be described as the rise of the mainstream creative-placemaking movement. Capturing this history is important as it allows one to see how certain ideas, actors, and locations have taken center stage in this story—and who has come to benefit from such a process. We trace this history from William H. Whyte in the 1950s, through the New Urbanists of the 1990s, and to Richard Florida in the twenty-first century, along the way paying attention to the roles of such prominent individuals and organizations as Jane Jacobs, Christopher Alexander, Richard Sennett, Ray Oldenburg, Robert Putnam, PPS, and the NEA, among others. What this history ultimately illustrates is that one cannot understand the evolution of the American city throughout the past sixty years without understanding the concept of placemaking. Contemporary placemakers have noted the influence of such actors and ideas on their work; their names appear in their public speeches and in the bibliographies of their white papers and reports. However, there has been little analysis of such influence and little work done to analyze and connect these often-disparate moments and influences. We seek to make these connections in order to achieve a better understanding of how these pasts inform the present state of creative placemaking in the United States.

By looking so closely at this broad history, we highlight how placemaking has, over time, consistently sought to address the tension between the individual and the community in the postwar city, primarily through an evolving understanding of the concept of sociability. As cities initially grew in both physical size and population during the post–World War II era, sociability, as fostered by placemaking, emerged as a tool to allow for the expression of both individual identity and communal belonging in urban centers where it was becoming harder and harder to find a balance between the two. Placemaking, in other words, became a way to hold a growing city together. And as these cities became more demographically diverse—and as corporate and government policies yielded catastrophic changes to the built environments of such cities—sociability became a salve for the personal and community wounds created by such intense transformations. By the 1980s, urban cultures, often expressed in terms of art and recreation, were at the center of a more productive brand of sociability. As more was on the line, placemaking had to do more to help cities and residents alike struggling to find identities, individual and collective, in the postindustrial climate of the late twentieth century.

By the early 1990s, placemaking and its emphasis on sociability was seen as leading to the production of what sociologists term social capital, which emphasizes the power of social networks to bring people together and create avenues of economic opportunity. At this time, well-known social scientists like Robert Putnam and lesser-known—but highly influential—academics like the historian Mark J. Stern and the urban planner Susan C. Seifert of the University of Pennsylvania's Social Impact of the Arts Project began to see the power of arts-based placemaking in cities hit particularly hard by deindustrialization and other forms of disinvestment. By the mid-1990s, these conclusions were echoed in the work of the sociologist Sharon Zukin, who argued in 1995 that, as industry left American urban centers, "culture is more and more the business of cities." By the late 1990s, Richard Florida would build on these efforts, suggesting that placemaking through art and culture attracted creative people to cities that embraced such methods, which in turn created environments where entrepreneurialism could flourish. Ultimately, this economic potential of placemaking proved irresistible in the postrecession moment: as the example of Gilbert attests, placemaking could lure cultural industries and the creative class to Detroit as the city sought to establish its postindustrial identity. Finally, such a process coincided with a scholarly rediscovery of the importance of the neighborhood, providing a setting—both spatial and intellectual—where the physicality of placemaking could be both grounded and put on full display. With the backing of such large-scale government agencies as the NEA and programs like ArtPlace America, placemaking as an economic-redevelopment tool had arrived.[19]

Yet the physicality of placemaking often highlighted only one understanding of such places while simultaneously allowing practitioners and observers alike to fixate on the surface components of this work and overlook the complicated histories and environments in which these projects came to exist. An honest assessment of contemporary placemaking must also take into account these oversights and omissions. Its one-size-fits-all approach overlooks nuances, and its attempt to act quickly leads it to elide such things as race- and class-based differences and inequalities. Its belief that it is doing something new leads it to be dismissive of what came before, while its emphasis on solving the postindustrial city leads it to emphasize certain populations that bring with them an affinity for certain professions and certain urban amenities. Many places and people have been left out of this growth. The current alignment of placemaking and traditional development ignores so much.

One sees how this plays out in Detroit, where Gilbert's work has yet to address the city's past or the histories of the city's predominantly African American neighborhoods. More specifically, Bedrock's downtown and midtown development plan does little to address the problems the majority of the city residents experience while channeling massive investment into the city's increasingly most valuable neighborhoods. This disparity is not lost on many Detroiters, who interpret the centralized development as being at least tone-deaf on race and civil rights issues, if not outright racist and classist. An instructive high-profile controversy emerged in July 2017 when graphics were being installed on the side of Bedrock's downtown Vinton Building. The window display was a photograph of an almost exclusively white street festival scene emblazoned with the slogan "SEE DETROIT LIKE WE DO." The social media storm that followed focused renewed attention on the development's inequities, forcing Gilbert himself to apologize in a Facebook post and admonish the marketing team: "Who cares how 'we see Detroit'?!" He continued: "What is important is that Detroit comes together as a city that is open, diverse, inclusive and is being redeveloped in a way that offers opportunities for all of its people and the expected numerous new residents that will flock to our energized, growing, job-producing town where grit, hard-work and brains meld together to raise the standard of living of all of its people." Yet the strongly

In 2010, a billboard celebrates Quicken Loans' move to downtown
Detroit over an empty Campus Martius plaza.

CityLoft, a pop-up holiday mini-mall from the suburban Troy's Somerset Collection mall, was an early retail experiment during Bedrock's transformation of downtown Detroit.

A mostly derelict block on Detroit's North Side.

worded post was about how "we screwed up badly the graphic package" and the complete installation would have been "very inclusive and diverse." Notably absent were statements of policy that might change more than the perception of a marketing program. Detroit city councilor André Spivey posted a response to the retraction: "I think Bedrock got the message. Let's see Detroit with everyone included."[20] But not all his constituents were sure anything but the message had changed.

A significant portion of this conflict is the feeling that downtown may not be for everyone. In the process of development, Bedrock has installed a massive surveillance apparatus, hiring hundreds of private security guards to patrol downtown, some unarmed, others armed. While security contractors are on the street, part of its in-house security force sits in a command center in the basement of Chase Tower reviewing video camera feeds from the firm's buildings. These are not necessarily the "eyes upon the street" of Jane Jacobs's imagination, nor are they the eyes of the full range of the city's residents. They may be watching the street, but who are they, and for whom are they looking?[21]

Gilbert's aim to "raise the standard of living of all of [Detroit's] people" is a laudable goal, but the underlying policies seem to involve making placemaking the cheery face of a redevelopment plan informed by trickle-down economics. And, while moving the Quicken Loans headquarters to downtown Detroit provides opportunities for white-collar employment, only 13.8 percent of working-age Detroiters have a bachelor's degree, and 21 percent do not have a high school diploma. Opportunities for Detroit residents are more likely to come from the other kinds of jobs created in downtown. Securitas, the security firm that Bedrock used for nearly five years, advertised jobs on CareerBuilder for as little as $10.00 an hour.[22]

Indeed, simply contrasting how Bedrock sees Detroit with the realities of the city reveals the shortcomings of a particular approach to placemaking. While Gilbert and others focus on attracting creative-class profession-als to downtown, the rest of the city remains starved for investment. And in an attempt to create a new narrative for Detroit—one apparently resting heavily on white residents—Gilbert and others ignore the geographies, histories, and needs of a number of predominantly low-income, African American and Latinx neighborhoods. Can any sort of placemaking speak to these communities? It is clear that the mainstream understanding of creative placemaking has not paid enough attention to such places. It is our contention that such spaces—such communities—highlight the need—and, indeed, the impetus for—another way to think about a more inclusive brand of placemaking.

"So That's Why We Fight": Another Take on 2010

But what would such an alternative approach look like? During the summer of 2010—at approximately the same moment that Gilbert was ramping up his efforts in Detroit and Markusen and Gadwa Nicodemus were writing their influential white paper—a project took shape in Chicago that was based on understandings of place and, ostensibly, could have fit into the category of creative placemaking. Yet this action did not take place at a conference, it did not feature academics and national organizations, and it was not celebrated by government officials. Instead, it seemed to offer a different model for such projects, one that highlighted race and rights while foregrounding different histories as it called for the production of something the twenty-first-century city no longer provided for in a community.

Faced with the toll that the recent financial crisis had taken on already diminishing education budgets, Chicago Public Schools (CPS) announced in early September 2010 the decision to tear down a field house it had designated as abandoned despite community members' insistence that they used the building—as they had been using it for decades—for community meetings and other activities. The field house was adjacent to Whittier Elementary School, a school located in the predominantly Latinx

South Side neighborhood of Pilsen. Parents initially lobbied for the site to be rebuilt as a library—a facility that the school itself lacked—but CPS spokesperson Monique Bond noted that demolition was the only viable option. "[With] the budgetary constraints we are under, nothing is going to happen," Bond explained. "That building has to come down."[23]

Before CPS could act, a group of parents, students, and sympathizers occupied the building, announcing that they planned to create a school library on their own. "We can't wait on CPS," proclaimed the Whittier parent Gema Gaete. "While CPS is twiddling their thumbs and being a stalemate and a spoiled brat, our children need these resources, so we figured we can do what CPS has been denying our children. We did it for ourselves." Soon, the structure was an even livelier center of community activity.[24]

Aesthetically speaking, the field house was not a pretty building; *The Wall Street Journal* described it as looking like "a long-deserted Pizza Hut." And some even refused to call the field house a building. CPS's Bond noted dismissively that it was "not even a building," just "a structure," and a structure on the verge of collapse at that. To combat such statements, Whittier parents hired an engineering firm to assess the two-thousand-square-foot building. On September 3, 2010, the firm Ingenii, LLC, issued a report concluding that the building was "in good condition and suitable for continued use." CPS rebutted these findings with a report by another engineering firm, Perry & Associates, that found that the building was not safe and that Igenii's conclusions were unfounded. For many of the parents now spending more time in the field house, such expert opinions mattered little: they knew the building was secure, and they still planned to occupy and use it as they saw fit.[25]

While the debate raged on outside the field house, the interior space was quickly renovated. It was painted with vibrant blues and greens, and a host of decorations transformed the space into a sort of inviting urban sanc-tuary. Commenting on life within the occupied building, a Whittier father reported: "Yesterday we had activities.... Guitars and dancing, tons of kids playing there, all kind of people were there." Such activity attracted the attention of even more parents, who—despite the presence of a large orange sticker on the front door proclaiming the building off-limits and threatening fines of $25.00 to $100 to all who entered—decided to take part in the reclamation of the space. No fines were ultimately levied, and no one was removed, despite the fact that the parents and their allies were technically trespassing.[26]

When asked why he decided to take part in the library campaign, the father cited above explained: "We are poor people. We're always losing because poor people don't mean anything.... So that's why we fight." The children at Whittier explained their involvement in similar language. A fifth-grader named Raul indicated: "I'm glad I get to go to school, but then I think about all the people who have libraries and we don't and then I remember that we're fighting for a library, for me and all the kids here who, like really, really, need a library." The budget cuts brought on by the Great Recession only made this need more acute. Another Whittier student reported: "We have only a few books in my class. Some classes don't have any books, and some classes got books that are so old and torn up that you can't almost read them." The roots of this crisis ran deep in Pilsen, where, by 2010, the poverty rate was close to 30 percent. The recent financial catastrophe served only to make these roots more visible.[27]

Such statistics make it clear that the Pilsen community had lived with the reality of dwindling resources for quite some time. Whittier once had a library, but the space had been turned into classrooms during the 1980s, and a similar story played out in over 150 CPS schools. Yet Whittier parents also understood that such policies of austerity could be challenged. Most recently, in 2001, parents in nearby Little Village set up a temporary camp at 31st and Kostner and went on a nineteen-day hunger strike, all to revive a forgotten promise for a new school in their high-density neighborhood. Eventually, the city made a $5 million down payment on what would become the Little Village Lawndale High School, which houses four high schools, including the Social Justice High School.[28]

In fact, this history of activism went back further than many people realized. Mexican migrants began to arrive in Pilsen in increasing numbers during the 1950s and 1960s. By the end of the 1960s, they constituted a majority of Pilsen residents. In 1954, the Pilsen Neighbors Community Council was founded to protest the effect of urban renewal on such neighborhoods as Pilsen. By the 1960s, murals depicting Mexican history were painted on buildings and viaducts throughout the neighborhood. In 1970, community activists succeeded in renaming and taking control of Howell House, a settlement house that had opened in 1905. Rechristened Casa Aztlan, it soon became a hub for political and community organizing.[29]

The field house was a part of this history—and had thus been a site of immigrant life for generations. Rose Escobar, a neighborhood resident in her fifties, remembered playing games, carving pumpkins at Halloween, and attending Christmas parties in the building throughout her lifetime. More recently, it had been used as a place where mothers of Whittier students gathered to take classes in sewing and English as a second language (ESL). For those who took part in such classes, the field house was more than a structure; it was a center of community.[30]

Motivated by these histories, the Whittier parents held a ribbon-cutting ceremony for their do-it-yourself (DIY) library on September 30, 2010. On the day the facility opened, sunlight streamed through the windows and yellow curtains and onto the crowded floor of the space. After a brief prayer, the schoolchildren raced into the building. There was still a ramshackle feel to the library; a wide assortment of children's books was stored in both milk crates as well as on attractive, multicolored bookshelves. Among the photographs and pieces of art that hung above these shelves was a poster featuring an image of a spaceship and the inspirational message "The

Sky Is the Limit." Bright rugs, tables, and chairs all came together to create an environment that was warm and inviting, a place that welcomed children of all ages. The space was undoubtedly cluttered, but the clutter removed the sterile feel often associated with school libraries. On the first day the library opened, it already felt comfortably broken in.[31]

Yet parents were quick to point out that their first task was simply to make sure the structure was not demolished. As one parent, Lisa Angonese, noted: "The grassroots [effort to create the library] came after we were fighting to keep the building. Remember, in the very beginning that was our main goal." As a single mother trying to raise two children, she found the field house to be a valuable resource for her family. And she drew on her own past as she undertook this effort: she had previously worked with the Pilsen Environmental Rights and Reform Organization (PERRO) in the past. PERRO, a grassroots community group that formed in Pilsen in 2004, sought to fight the disproportionate amount of pollution in the neighborhood. Angonese herself worked to close two nearby coal-fired power plants, the Fisk and the Crawford Generating Stations.[32]

Angonese felt, however, that, once occupied and spared the wrecking ball, the field house quickly became a place full of life and activity. She explained: "We held after-school tutoring with students from Roosevelt University, and we had parties, workshops, and, when the library was put together, we had author visits, art classes, book readings, and the kids were allowed to check out books. It was a lending library for the community." This lending library was supported by a host of established local institutions, including the Chicago Teachers Union, which, in addition to vocally supporting the activists, donated five hundred books to the facility. Other donations arrived from a variety of public and private actors, including schools and individuals around the nation, from such states as New York, Minnesota, and Louisiana. But the project also enjoyed the support of new organizations in the city of Chicago, including the Chicago Underground Library, a group founded in 2006 whose slogan at the time was "Slow media, locally grown, pesticide-free."[33]

The Chicago Underground Library (which changed its name to Read/Write Library Chicago in 2011 and has no relationship with the city's official library system) put out a citywide call for donations and helped organize the books coming in. At the time of the Whittier struggle, it saw itself as "a new model for open, location-specific archiving of independent and small press media." It sought to acquire, catalog, and make available books, magazines, zines, journals, newspapers, and art books of all genres from the Chicago area, seeing its commitment to "obsessive cataloging" as a way to map the evolution of the city's communities and movements over time. At Whittier, the group saw real grassroots belief in the importance of libraries and was, according to Chicago Underground Library executive director Nell Taylor, able to mobilize a host of "unemployed and underemployed librarians" to work with the parents in creating a functioning library. Such librarians, she noted, like the schools they once served, had been hit hard by the financial crisis: "City, state, and federal agencies are using the crisis as an excuse to cut the things [out of budgets] that they have wanted to cut for a while." For Whittier parents like Angonese, the efforts of Taylor and others had helped them create a space, a resource, that the postindustrial city had taken from them. And, despite the attempt to make that space as attractive and appealing as possible, Angonese was quick to note: "We did not do any type of creative placemaking at all."[34]

On October 19, 2010, CPS CEO Ron Huberman, who had become public enemy number one among many Whittier parents, announced that the field house would not be razed—and that CPS would use the funding allotted for its demolition to renovate it. School officials also pledged to build a library inside the school and lease out the field house to Whittier parents for a fee of $1.00 per year. On October 28, the parents had the deal in writing. On November 3, Huberman announced that he was step-

ping down as the head of CPS. Parents quickly began collaborating with a team of architects, including Raymond Barberousse, Robert Quellos, and Katherine Darnstadt, on the design for the new building. These professionals, working through an organization called Architecture for Humanity (which changed its name to Open Architecture Collaborative in 2016), held biweekly meetings with Whittier residents as the design process proceeded. "The parents," noted Darnstadt, "were with us every step of the way." The Whittier activists wanted a multifunctional space, one that could hold a library while having room for ESL classes and after-school programs. But, having lived with industrial pollution for so long, the parents also wanted an environmentally sustainable building—one that could become a model for other schools throughout Chicago. In explaining the decision to strive for Leadership in Energy and Environmental Design (LEED)–certified status, Darnstadt explained that the design group was motivated by the question, "Can we design a building that could be a prototype for how sustainable a school building could be?"[35]

Sadly, this question was never answered; the 2011 election of Rahm Emanuel stopped such plans from becoming reality. Campaigning for mayor, Emanuel made it clear that he saw reforming CPS as one of his major goals once in office. Such a position made the renovation of La Casita a nonstarter. Instead, the new CPS CEO, Jean-Claude Brizard, called for the creation of a library within Whittier Elementary itself. Afraid that such an arrangement would displace special-education students in an already overcrowded school, parents rallied against this plan. For close to two years, a stalemate of sorts held. Yet, on August 16, 2013, that stalemate was broken when CPS issued a statement that the building had to come down because it contained asbestos. This was the first time during the three-year ordeal that the presence of this hazardous material within the field house had been reported. On August 17, 2013, La Casita was demolished under orders from CPS.[36]

La Casita's opening-day
celebration in 2010.

As the Whittier parent Lisa Angonese noted, no one involved with La Casita explicitly described what the parents were doing as *creative placemaking*. Yet it is our contention that one sees the seeds of a potential alternative approach to mainstream creative-placemaking efforts within this effort—and the second half of this book will use other similar projects as the means of mapping out the characteristics of this approach. As Whittier highlights, such work privileges people, neighborhoods, histories, and physical spaces often left out of mainstream placemaking projects. Attuned to issues of race and class in ways that highlight how histories associated with civil rights, subaltern urban cultures (in the shape of, e.g., hip-hop) and activist groups can come to inform an alternative approach to placemaking.[37]

But, most importantly, these projects create sites of production. As mainstream creative placemaking valorizes the entrepreneurial potential of the postindustrial economy and the role of the creative class in such projects, we highlight places where people left out of those narratives are actually making things. On the one hand, these can be seen as acts of necessity: urban residents simply creating goods and services—like a library—that the neoliberal city no longer provides for them. But we see it as something more: a road map for a more just, equitable city for individuals and communities alike.

Of course, these struggles do not always achieve what they set out to accomplish. As we see at Whittier, not only is La Casita gone; the school still does not have a library. It is important to reckon with the institutional and structural factors that frustrate these efforts. For Whittier, changing institutional leadership and the expansion of a neoliberal approach to public education undermined much of the concrete progress the group had made during its occupation of the field house. But, even in such cases, we need to find new ways to understand influence. The same goes for actions that are intended to be temporary. After all, for both individuals and organizations, involvement in such projects can strengthen resolve and push in new direc-

tions. The Whittier parent Lisa Angonese continued to work with PERRO, fighting to get both nearby coal plants decommissioned in 2012. She then went on to serve as a peacekeeper for the marches and demonstrations that greeted the 2012 NATO summit in Chicago and defended the "NATO 5," a group of activists arrested for alleged criminal activity before the summit. Speaking on how La Casita influenced these actions, Angonese offered: "The experience helped me to learn and stand with others on many different platforms."[38]

Another Whittier activist, Evelyn Santos, was similarly motivated by her participation, serving two years in Public Allies Chicago, an AmeriCorps program that works toward social justice by "engaging and activating" young people. "I wanted to continue helping other communities and trying my best to empower other people," Santos said. She also noted that the attention generated by La Casita did bring some changes to Whittier Elementary School; CPS was compelled to update some of the school's aging infrastructure and install a small computer lab.[39]

For Read/Write Library Chicago, the experience led to a more place-based approach anchored in the specific communities with which it works. While students and others from across the city visit its main location in Humboldt Park, the group's interest in developing collections within communities regularly brings it into other neighborhoods. The group formally implemented a citywide pop-up library program whose collection is not only tailored to the place where it is located but also created by the library's users as contributors. Read/Write also operates a miniversion of these pop-ups, BiblioTreka, a traveling library that moves throughout Chicago attached to a bicycle. But support it offered during the Whittier experience also clarified something essential for the group. While the Chicago Underground Library's name and spirit comported with its own initial identity, Taylor came to understand that the underground nature of the Whittier project was not "a reclaimed identity or a rebellious choice to operate outside a system. . . . It was underground be-

In 2013, a banner decries the demolition of the Whittier field house, with the former community center's rubble heaped behind the temporary fence and the police presence.

cause it was devalued by the people who had power to shape what the contemporary landscape looked like, and, through education policy, had the power to shape what histories and culture were given visibility."[40] Even though La Casita's library was underground, identifying it as such did not match the activists' aspirations. The parents agitated for the library to be fully sanctioned and supported by the city—anything but underground. The realization was a push toward a new name that recognized the democratic, responsive, and grassroots nature of the organization and its work.

For the architect Katherine Darnstadt, the episode provided a lesson in "how institutions can break down community activism." Yet working with Whittier activists also provided her with an understanding of how democratic, place-based design could ultimately serve an urban population. In 2015, her firm, Latent Design, initiated the Boombox program. Through Boombox, entrepreneurs and small-business owners can rent out space on a sliding scale in a pop-up, prefabricated micro retail kiosk. These kiosks provide space for those who do not have access to the capital necessary to buy a building or sign a long-term lease; the program's slogan is "Between startup and storefront." While the first Boombox was in hip Wicker Park (and featured Read/Write Library Chicago as a tenant), successive pop-up storefronts have been in Austin, Chatham, and Englewood, three Chicago communities usually overlooked when it comes to such programming. Previous tenants include Justice of the Pies (a piemaker), Mod+Ethico (a clothing retailer), and Shyne City (a shoe-shining business), among others, each highlighting unique urban histories and cultures while showing that such places can still make things. "Without Whittier," concluded Darnstadt, "I would have never taken on such a project."[41]

Shifting attention from Dan Gilbert's downtown Detroit and toward Pilsen and other underserved and overlooked neighborhoods of Chicago reveals the complexity

(*Above and following spread*) Shyne City at a Boombox installation
in Chicago's Chatham neighborhood in 2018.

missing from the traditional narrative about creative placemaking. The parents who created La Casita produced cascading personal, organizational, and other changes from both a position and a history outside the traditional power structure. These efforts remind us how the broad expanse that constitutes creative placemaking has included some actions at the expense of others. After all, the occupation of a field house by Latinx parents and their supporters falls into Markusen and Gadwa Nicodemus's definition of *creative placemaking* as much as Detroit's Campus Martius does, with the activists' emphasis on animating public and private spaces, rejuvenating structures, and of course doing far more than simply fostering entrepreneurs and cultural industries—but really doing that, too. Moreover, the fact that Whittier participants do not see their work as creative placemaking demonstrates how the concept is deployed today and what and whom it has come to represent.

This book, then, is no antiplacemaking screed. The ideas and projects developed and advanced by mainstream placemakers like Gilbert and ArtPlace America have begun valuable conversations about what the twenty-first-century city should look like and how it should operate. And we do not want to jettison the belief that good urban places should enhance sociability. After all, even though some of the best-known advocates of such an idea—including Jane Jacobs and Robert Putnam—argued that sociability is important, they also saw it as an important step in a much bigger process. Despite this recognition, the creative-placemaking discourse and related policies have overemphasized sociability and other such stages of placemaking (e.g., the perceived economic benefit that flows from increasing sociability), limiting the potential of both individuals and communities. We must pay attention to where sociability is directed and why.

The second half of this book documents groups and efforts that do just that. We have spent ten years searching for and connecting with such people in community gardens, record stores, and bars, during community meet-ings, protests, and conferences, and, of course, over social media, on Internet message boards, and more. We have visited with more than two hundred artists, urban farmers, dancers, activists, entrepreneurs, and others in more than forty cities throughout the United States who are not satisfied with simply activating a place. Through these visits, we have had scores of brief but influential discussions, developed close friendships, and become long-term collaborators. We have also conducted phone and email interviews with such individuals and consulted public statements made by them. They—and we—see activation as just the beginning. We could not include all these interactions in this book, but we highlight dozens of cases in the text and include even more in photographs to highlight just how widespread these practices have become. As white, middle-class academics, we may not fully discern certain aspects of these projects. But we wish to start by demonstrating how, for these actors, placemaking is more than a feel-good marketing ploy or a way to get the right people to hang out together; it is a way to begin transforming the social as well as economic, political, and cultural forces that have given us the city of the early twenty-first century.

These conversations can, as one might expect, be highly critical of what mainstream placemakers like Dan Gilbert choose to celebrate. Here, the shadows of gentrification and displacement loom large. Yet such voices also call for work that provides a more positive way to rethink the relationships between place and the urban environment. Most importantly, as we expand the discourse and explore examples outside the creative-placemaking canon, we note that the larger aim of these projects is often production—and how *production* is defined is dependent on the environment in which actors labor. While great attention has been paid to the knowledge- and arts-based labor of creative workers within mainstream placemaking endeavors, we stress that thoroughly unpacking placemaking processes requires critically assessing the relationship between tangible and intangible outcomes.

After all, physicality is often at the core of production and placemaking, with such tangible things as food, housing, and jobs being rooted in a variety of approaches to work and labor. These production-based outcomes take account of the influence and efforts of actors, organizations, and histories commonly overlooked by mainstream placemaking practitioners. By allowing these counter-narratives to take center stage, we hope to democratize this burgeoning field of creative placemaking.

In calling for such a deeper examination of placemaking, we acknowledge the economic and social practices that define productive forces, but we promote a conception of production that extends far beyond the marketplace. We want to highlight directed action and its products, whatever they may be. Here, the intangible outcomes often associated with traditional placemaking endeavors, like new relationships or community identity, are still produced. More specifically, such a holistic understanding of production emphasizes the value of bonding and bridging social capital—the ideas that communities derive benefits from connecting within and between each other. But that sociability can also be directed toward the production—and subsequent distribution—of something more. This potential relationship between intangible and tangible productive outcomes can be seen in the example of La Casita, where activists' efforts bolstered and broadened a community, transformed professional practices, and inspired future action by temporarily transforming a field house into a brick-and-mortar library. Creating and maintaining this physical resource for students and community members was essential to the struggle. Moreover, as in so many of the cases we outline in this text, the La Casita activists produced new interpersonal and organizational pathways for the redistribution of other resources. In the short term, these pathways redistributed books, volunteers, and money from around the country to a group of concerned parents on Chicago's South Side; the long-term effects of these networks

are still being authored. Creative-placemaking projects are not often so productive. Not being specific about the means and ends of projects can lead to such work falling flat and even reinforce existing and often unequal social dynamics. Being clear about what—and for whom—creative placemaking produces—and how those gains are distributed—is part of being clear about what it could be. Placemaking must, in other words, be about more than the recruitment of the creative class; it must speak to the needs of those indigenous to the communities in which these projects are rooted. It must embrace a vision of production that is holistic, redistributive, and cognizant of the connection between tangible and intangible outcomes.

Such clarity means understanding that even alternative ways of practicing creative placemaking can lead to unequal economic development and displacement. In certain situations, the success of such projects can lead to outcomes that can harm those whom the project was initially designed to assist. At the same time, these alternative approaches can lead to projects that are less than permanent by design—or that strive for permanence and ultimately fail to last. The temporary aspect of many of the projects we document in this book must be considered in any honest assessment of such work.

Ultimately, then, while we hope that the case studies, histories, and discourses analyzed here illuminate key features of what creative placemaking is and could be, the goal is not to produce a comprehensive history of creative placemaking as a contemporary practice, nor is it to provide a how-to guide for creative placemaking. Instead, we strive to use the processes of embedded observation and community-rooted engagement to help illuminate the merits, demerits, and possibilities of a more inclusive, more equitable approach to placed-based action. Each history provides its own lessons for action and ways of seeing—if we can look past the ways we have learned to interpret what placemaking is and is not.

With all this in mind, we now turn our attention to unpacking the long history behind contemporary placemaking. We wish to understand the development of its ideas and ruminate on the actors, influences, and cases that have taken up so much of the attention given to this topic.

Such a strategy does more than lay out a valuable history behind a contemporary phenomenon; a comprehensive understanding of the evolution of mainstream creative placemaking will allow us to seek out and develop alternative approaches more effectively.

The (Near) Death and Life of Postwar American Cities

The Roots of Contemporary Placemaking

(*Previous spread*) Bryant Park, an early site of creative placemaking, on a spring evening.

It often appears that the placemaking movement does not have much of a past. It rose rapidly in the early twenty-first century to deal with problems of the here and now, and history understandably has taken a back seat. Moreover, placemaking by definition is a forward-looking process, working to transform a particular past while inviting observers to imagine what the place *could* be. Yet, as laid out in this book's introduction, it is clear that contemporary advocates of placemaking are self-consciously drawing from a specific body of literature and a certain history of urbanism within the United States.

This chapter seeks to position twenty-first-century placemaking efforts as part of this lineage. More importantly, we place contemporary placemakers in dialogue with this history as, in their formal reports and informal writings, these placemakers place themselves. Here, we offer an intellectual history of contemporary placemaking as we document the diverse body of influences—including people, documents, and ideas—informing the ascendant, mainstream version of twenty-first-century placemaking as practiced by such actors as government leaders, philanthropic foundations, nonprofit organizations, and arts-based groups. To document all these influences would be impossible. Instead, we focus on those texts that have had the most pronounced sway over the evolution of placemaking. All the sources highlighted in this chapter crop up in myriad reports, how-to guides, and academic works on the topic of contemporary placemaking in the United States, constituting the first coherent core of readings orienting the development of placemaking.

Yet contemporary placemaking's relationship with this past is more than an intellectual one. Placemakers turn to this history for practical advice as it is within this body of influences that one sees the roots of the key concepts that help define such contemporary placemaking efforts, including an emphasis on the neighborhood as the prime unit of organization, a turn away from a manufacturing-based economy and toward the cultural creative economy, the benefit of informal street-based surveillance,

and the belief in organic redevelopment that eschews one-size-fits-all planning strategies. Ultimately, however, these influences were most interested in creating spaces of sociability that balanced the needs of both individual and community. These new spaces could have significant economic and political consequences, but those outcomes often took a backseat to personal development. Such a focus has compelled contemporary placemakers to overlook a host of movements, individuals, and ideas that do not fit in with this approach.

An astute reader may observe that we offer a selective reading of the sources we highlight as integral to the intellectual development of contemporary placemaking. This is by design. We are interested in exploring the ways placemakers themselves read—and engage with—these texts. Their process is anything but complete, and our approach mirrors this reality. Placemakers embrace an ad hoc approach to this past, one that is even contradictory at times. The attitude overemphasizes certain aspects of the sources while underexamining others. In short, contemporary placemakers are often too narrow in their interpretation of essential readings and ignorant of the limitations acknowledged by the texts' own authors. The resulting conception of placemaking is thus thin in critical ways. We will point out these moments as they occur as they will be instrumental in later chapters.

The Limits of the Organization Man: William H. Whyte and the Case against the American Suburb

To understand the rise of twenty-first-century creative placemaking—a phenomenon primarily rooted in cities across the United States—one must first grapple with the rise of the post–World War II American suburb. For, as this type of living arrangement came to dominate American culture, there was a sense that the places that such suburbs seemed destined to replace—the nation's cities—were a remnant of the country's past. By 1950, the national

In Camden, New Jersey, the kind of landscape often eliminated during urban renewal.

growth rate for America's suburbs was ten times that of central cities. In 1954, *Fortune* magazine estimated that nine million people had moved to the suburbs during the previous decade. And there was no doubt from where these people were coming. Of the country's twenty-five largest cities in 1950, eighteen lost population over the following three decades.[1]

As the work of scholars such as Elaine Tyler May has shown, the postwar American suburb was deemed by policymakers and cultural commentators alike as a necessary bulwark against the insecurities ushered in by the burgeoning Cold War and the memories of both the

Great Depression and World War II. Yet, despite the order and stability that such places provided, some came to bemoan the loss of urban centers and the apparent damage wrought by the rise of the suburb on the individual psyche of countless Americans. Works such as David Riesman's *The Lonely Crowd* (1950) and Robert A. Nisbet's *The Quest for Community* (1953) expressed early reservations regarding the toll that the new era was taking on the internal as well as external lives of countless Americans. By 1957, in *America as a Civilization*, Max Lerner looked to the suburb and found that "the problem of place in America" was now the nation's most pressing issue. If this problem were not resolved, Lerner concluded, "American life will become more jangled and fragmented than it is, and American personality will continue to be unquiet and unfulfilled."[2]

Such anxiety undoubtedly informed the strategies associated with urban renewal. Admitting that the city of the postwar era was on the cusp of perceived obsolescence, urban planners sought to erase the urban fabric as it existed and revive landscapes that were seen as dirty, outdated, and unsafe. Urban renewal offered a new vision of how a city should look and work, a reconceptualization of what a city could be. Cities as disparate as New York, Houston, New Haven, and Milwaukee all came to embrace the logic of modern urban renewal in the early postwar era.[3]

Owing to the work of a number of scholars, we have a strong understanding of how this logic came to pass. And, while urban renewal was often prescribed and carried out with good intentions, it ultimately proved disastrous to the well-being of the cities it was attempting to save. Yet there were those who chafed at the stifling realities of the American suburb while not advocating the wholesale destruction that was urban renewal. As early as the 1950s, a counternarrative of sorts emerged, one that

Prairie Shores, a major urban renewal–era development on Chicago's South Side.

argued against the imperative of suburbanization while calling attention to what a reconfigured city could once again provide. This was the start of a decades-long conversation that would come to inform placemaking in the early twenty-first century.[4]

This counternarrative must begin with William H. Whyte, whose shadow looms large over this entire discussion and who can best be described as the founding father of placemaking in the United States. His work continues to be studied by contemporary placemaking advocates and cited in reports on the topic issued by such bodies as the National Endowment for the Arts (NEA). Moreover, the group he founded in 1975—Project for Public Spaces (PPS)—remains a key player in the realm of placemaking, active in cities around the world. Throughout the postwar period, Whyte remained an ardent believer in the usefulness—indeed, the necessity—of the American city. Nevertheless, the urban landscape that he saw as ideal looked very little like the city proposed by urban renewal advocates. His approach to urbanism and the ways that it fostered relationships between individuality, sociability, and even economy influenced those who followed him—and continues to influence the field of placemaking in the early twenty-first century.

Interestingly, the book that made Whyte a household name was not about the city; it focused instead on the American suburb. In 1956, when he penned *The Organization Man*, the American city had seemingly been eclipsed by the suburb. Widespread, large-scale urban renewal efforts suggested that the city was broken. And Americans were seemingly voting with their feet, moving to the burgeoning suburbs by the millions. To Whyte, the opportunities afforded by an increasingly corporatized economy were making such movements possible. Those working in these large-scale corporations—predominantly young white men—saw in the American suburb a landscape that seemed to complement their professional lives.

Whyte was none too pleased by this development. Surveying the demographics of the growing American suburb, he saw what he dismayingly deemed "a gener-

ation of bureaucrats." As he reported in *The Organization Man*, he was struck by the cautious nature of the nation's graduating college seniors. Such young people valued a life that was "calm and ordered," one preferably lived within the security of a large corporation. Whereas previous generations sought to strike out on their own, the college seniors of the 1950s "do not bother with this sort of talk": "Once the tie has been established with the big company, they believe, they will not switch to a small one. . . . The relationship is to be for keeps." Not surprisingly, these young people tended to "shy from the idea of being entrepreneurs"; individuality and creativity were cast as roadblocks on the path to postwar affluence.[5]

Yet what was most striking about Whyte's account of these burgeoning bureaucrats was his description of the places they came to call home: the American suburbs. His treatment of the places of the organization men and their families privileged both the newness and the uncertainty of such spaces. Here, the young corporate workers were cast as "the transients," a group marked by a sense of "rootlessness" that was reflected in the spatial organization of the suburb. "In suburbia," as Whyte found it, "organization man is trying, quite consciously, to develop a new kind of roots to replace what he left behind."[6]

To Whyte, the American suburb as it emerged in the 1940s and 1950s was "a hotbed of Participation," a place that had quickly developed "a communal way of life." Yet at the same time he also noted how these suburb dwellers were simultaneously quite aware of the impermanence of such spaces; fear of rapid turnover in suburban homes and rental units, for example, was on the minds of many such individuals as these new workers could be moved at the whim of the organization. Such realities meant that, "psychologically, at least, the newcomers to suburbia are living on the brink of a precipice."[7]

It was up to the suburbanites to navigate such conditions, and they did this by developing what Whyte termed the *social ethic*. As he put it: "The transients' defense against rootlessness . . . is to get involved in meaningful activity; at the same time, however, like the seasoned

Houses built on the edge of Cleveland in the post–World War II period.

shipboard traveler, the wisest transients don't get too involved. Keeping this delicate balance requires a very highly developed social skill, and also a good bit of experience."[8] On the one hand, such an ethic had little room for individualism as the "quest for normalcy"—which Whyte ultimately believed was "self-destructive" and "one of the great breeders of neuroses"—was the core component of the communal spaces of the suburb. Yet the sense of community offered by such places, which was stressed as their prime selling, was also less than satisfactory, for it remained both remarkably shallow and often transitory. For the average suburb dweller, then, these spaces were the worst of all worlds.[9]

But these spaces appeased the American political economy of the mid-twentieth century as they created an environment that fostered the character and temperament necessary to maintain corporate order. "As far as social values are concerned," Whyte concluded, "suburbia is the ultimate expression of the interchangeability so sought by organization." Here in suburbia were the breeding grounds for workers who would spend their lives within the bureaucracy of the large-scale corporation. The suburbs taught such workers to get along with each other—but not to get too close—while dampening any sort of individual initiative or thought.[10]

For Whyte, not only would such environments soon breed neurotic people; they would also create a business model unable to innovate and evolve over time. In light of the conformity encouraged by the American suburb, he argued: "What we need is not to return but to reinterpret, to apply to our problems the basic idea of individualism, not ancient particulars." The corporate world must embrace the "encouragement of individualism" and the belief that "the individual is more creative than the group." And, while in the mid-1950s Whyte did not call for the end of the suburb, he did see it as a less-than-stellar place to begin to fight "The Organization." Not surprisingly, his writings would soon come to focus on another American space, one perhaps more likely to encourage the sort of individuality that he thought the nation needed to grow: the city.[11]

Here Comes the Neighborhood: Jane Jacobs and the Revival of the American City

Helping shape Whyte's thinking on the American city was an up-and-coming writer who was beginning to take on government-sponsored urban renewal efforts. It is not stretching things to refer to Jane Jacobs as the patron saint of contemporary placemaking practitioners; her 1961 *The Death of Life of Great American Cities* has become something of a bible to the movement. In the face of the monotony and soul-crushing sameness of urban renewal strategies, Jacobs raged against the "great blight of dullness" that had come to define the mid-twentieth-century American city. In light of such sterile environments, she wrote of the value of urban "eye-catchers," spaces that provided color to their drab surroundings. "Some eye-catchers," she found, "are eye-catchers just by virtue of *what* they are.… Other eye-catchers, however, are eye-catchers because of precisely *where* they are." Both types were valuable to a city's built environment.[12]

These spaces put on display the vitality of urban life, with the focus on seeing the city proving quite telling. On the one hand, the visibility of such spaces provided the cues to city residents that they were, indeed, places of sociability meant to be utilized and enjoyed. Yet the benefits of visibility also extended into the realm of urban surveillance. To Jacobs, "there must be eyes upon the street" to bring about "the self-government functions of city streets: to weave webs of public surveillance and thus to protect strangers as well as themselves; to grow networks of small-scale, everyday public life and thus of trust and social control, and to help assimilate children into reasonably responsible and tolerant city life."[13] With these ideas in mind, public spaces such as sidewalks and streets thus became more than mechanisms of transportation, of movement. They became places of both safety and sociability. The state, it should be noted, seemed to play little to no role in meeting such responsibilities.

For Jacobs, the fact that a street could do so much was evidence of the necessity of multiuse spaces within cities. Single-use spaces, often maintained through deliberate zoning restrictions, worked only to create dead zones, or places where people did not want to congregate. Public housing in American cities, with its rejection of mixed-use planning, provided ample evidence for Jacobs here. To counter such damaging social isolation, she noted: "Effective diversity of use, drawing deliberately a sequence of diversified users, must be deliberately introduced [into public urban spaces]." This meant that the city should be working to create spaces that hosted such diverse activities as ice-skating, washing bikes, roasting pigs (!), and street vending. Such "deliberate street arrangements for vendors can be full of life, attraction and interest," Jacobs argued, and would bring together a remarkable mix of disparate city dwellers.[14]

Yet Jacobs made it clear that such mixed-use spaces did not just crop up. Instead, they needed what she referred to as "an enormous diversity of ingredients," integrated to create a sort of formula for successful urban spaces. For starters, she found that the district in question must serve more than one primary function. Second, most blocks within the area had to be short ("that is, streets and opportunities to turn corners must be frequent"). A commitment to maintaining buildings of all sorts and ages also had to be fostered, as did a "sufficiently dense concentration of people." All these ingredients were necessary for the cultivation of a successful public space—and the absence of even one could be highly damaging to the locale's overall usefulness.[15]

Such a focus on the cultivation of exemplary urban spaces did not mean that Jacobs was an advocate of one-size-fits-all urban planning. Instead, she believed that urban development had to be organic, coming from the ground up, and seeing spaces as part of the specific fabric of their immediate surroundings. In fact, she stressed that it was counterproductive to think of such efforts as part of any sort of larger program or even to see them as projects in and of themselves. Instead, the goal should be to get "that patch of the city . . . rewoven back into the fabric—and in the process of doing so, strengthen the surrounding fabric too." By calling attention to the context of spaces in question, Jacobs came to see how the benefits of existing structures, for example, could be recast and harnessed anew to rethink a particular urban space. Within such a mindset, old buildings, previously seen as blight, were recast as spaces of opportunity. And Jacobs was not simply advocating for the use of museum-piece old buildings, though such structures could be culturally significant and therefore useful to a community's overall health. She also saw great value in maintaining "a good lot of plain, ordinary, low-value old buildings, including some rundown old buildings." These buildings, with their low cost of entry, would allow all sorts of enterprises to open for business.[16]

Here, one is struck not only by Jacobs's adoption of the strategy of adaptive reuse—old buildings being used for entirely new purposes—but also by her insistence that such urban spaces should be thought of as the logical birthplaces for new entrepreneurial endeavors. It was in such repurposed spaces that small-scale start-up firms could plant their roots. "Wherever lively and popular parts of cities are found," Jacob found, "the small [enterprises] much outnumber the large." Perhaps more importantly, these spaces allowed for creativity and innovation to be developed by those often priced out of big cities. Such environments, Jacobs concluded, "*are* natural generators of diversity and prolific incubators of new enterprises and ideas of all kinds."[17]

Of course, what many readers culled from such a robust discussion of urban life could be boiled down to one phrase: *neighborhoods matter*. And there is some truth to such a reading of *The Death and Life of Great American Cities*. Jacobs does, after all, devote an entire chapter to the topic of the uses of city neighborhoods. Most broadly, she argued that city neighborhoods were, simply put, a necessity for urban living. As she wrote: "But for all the innate extroversion of city neighborhoods, it fails to follow that city people can therefore get along magically without neighborhoods. Even the most urbane citizen does care about the atmosphere of the street and district where he lives, no matter how much choice he has of pursuits outside of it; and the common run of city people do depend greatly on their neighborhoods for the kind of everyday lives they lead." City neighborhoods thus emerged as "organs of self-government."[18]

Yet Jacobs offered a far more nuanced understanding of neighborhood than many of her acolytes have noted. To her mind, too many planners had tried to use the concept of neighborhood to replicate the social conditions of the early American town as well as the then-burgeoning suburb. Because of its sheer size, however, the mid-twentieth-century American city had little in common with either of these spaces. In light of such realities, Jacobs concluded: "Neighborhood is a word that has come to sound like a Valentine. As a sentimental concept . . . [i]t leads to attempts at warping city life into imitations of town or suburban life. Sentimentality plays with sweet intentions in place of good sense."[19]

At the same time, Jacobs also offered a further explication of the concept of city neighborhood itself. There were, in fact, three kinds of such neighborhoods: "street neighborhoods"; "districts of large, subcity size, composed of 100,000 or more in the case of the largest cities"; and "the city as a whole." Rather than seeing the most value in the micro-level street neighborhood (which did provide the self-surveillance necessary for the well-being of an urban community), in arguing for the primacy of seeing the city as whole Jacobs found that "a city's very wholeness in bringing together people with communities of interest is one of its greatest assets, possibly the greatest." It was then the job of the district to mediate between the "indispensable, but inherently politically powerless," street neighborhood and the powerful city as whole, helping distribute the resources offered by the latter to the residents of the former. Such a job suggested a size for districts larger than the standard planning definition of *neighborhood*. According to Jacobs: "An effective district has to be large enough to count as a force in the life of the city as a whole. The 'ideal' neighborhood of planning theory is useless for such a role." Such nuance would come to be overlooked in later assessments of her work—and by those within the contemporary-placemaking movement who look to her work for guidance.[20]

Jacobs's next book, 1969's *The Economy of Cities*, immediately picked up with the admission that her previous work had illustrated that cities are often understood as "primary organs of cultural development." She now wished to make the case that "cities are also primary economic organs."[21] Much had changed in the eight years since the publication of *The Death and Life of Great American Cities*, and Jacobs implicitly realized that she was writing at the very moment that the urban crisis was gripping with cities like her beloved New York. What was helping feed this moment of crisis? To Jacobs, it was a fundamental misreading of the evolving urban economy. Policymakers and business leaders alike continued to believe that large organizations could still grow a city's economy. Such organizations had, like large-scale urban renewal plans, proved ineffective. "The infertility of large organizations is not a new phenomenon," Jacobs found. "Large economic organizations are seldom able—once they have become large—to continue adding enough new activities to keep themselves from shrinking."[22]

More specifically, Jacobs saw a continued reliance on the perceived economic benefits of large-scale manufacturing as a grave mistake. Such an economic strategy was, she concluded, one best left in the past. "Mass-production manufacturing," she noted, "will no longer by regarded as city work…. Manufacturing work will, I think, no longer be the chief activity around which other economic activities are organized, as it is today and as the work of merchants once was." So what was the future for urban economies? To Jacobs, "services will become the predominant organizational work" as these types of work will become "the instigators of other economic activities."[23]

Such a future seemingly had little place for the urban working class and the institutions that represented their interests. Labor unions simply fought to protect the status quo, stifling needed innovation and flexibility. At the same time, such bodies attempted to paper over the fundamental disorganization of those who populated them. "The inherent solidarity of the working class," Jacobs wrote, "is an economic fiction."[24]

A similar check on economic innovation was also seen in government policy. Within her native New York, for example, Jacobs came to see economic stagnation as due in no small part to the "remarkable growth of unproductive make-work in the city bureaucracies." More importantly, the government's lack of creativity and inability to think small were also issues. The question of "why governments should be so imitative, rather than innovative," vexed Jacobs, as did the belief that government leaders "[did] not seem to bring their minds to bear on a particular and often seemingly small problem in one particular place." That was "how innovations of any sort are apt to begin," but governments still tended "to seek sweeping answers to problems; that is, answers capable of being applied wholesale the instant they are adopted." This had not worked with urban renewal; it would not help jumpstart the economies of American cities.[25]

And what would such an approach to urban economic development actually look like? For Jacobs, the late twentieth century would see plenty of "room for extraordinary growth of learning and the arts in the large local econo-mies of big cities."[26] The individuals working within such fields shared a similar trait that Jacobs came to see as vital to the economic health of the American city: creativity. It was up to the city to find ways to allow this creativity to flourish, a process that would mean leaving some economic actors and activities in the past. Cities, in other words, had to "upset the status quo, make some well-established activities obsolete and reduce the relative importance of others." What was playing out was nothing less than a battle for the future of the urban economy. The "primary economic conflict" had become, Jacobs concluded, one "between people whose interests are with already well-established economic activities, and those whose interests are with the emergence of new economic activities." Here are the beginnings of the creative class, that great force for disruption within American cities that the urban theorist Richard Florida would chronicle over thirty years later.[27]

By 1984, Jacobs had grown worried that cities were not heeding such advice. That year's *Cities and the Wealth of Nations: Principles of Economic Life* began with the observation that too many cities remained wedded to such "well-established economic activities" for far too long. "The country's manufacturing economy has gradually but steadily been eroding," Jacobs wrote, "and much of what remains has been slipping into technological backwardness relative to industry in Japan and the more vigorous parts of Europe."[28] Much of this was, she felt, due to the fact that economic activity remained inextricably linked to the concept of nation. This was a mistake. "Nations are political and military entities," she noted. "But it doesn't necessarily follow from this that they are also the basic, salient entities of economic life." In fact, this belief in the nation as the primary unit of economic activity had actually seemed to hinder such activity. "Indeed," she concluded, "the failure of national governments and blocs of nations to force economic life to do their bidding suggests some sort of essential irrelevance."[29]

In light of such perceived irrelevance, it was time, Jacobs posited, to assert the primacy of the city as economic catalyst: "Cities are unique in their abilities to shape and reshape the economies of other settlements, including those far removed from them geographically."[30] Perhaps more importantly, it was at the city level where individuals could successfully circumvent the limitations of earlier ideas of urban economic development. By the mid-1980s, Jacobs had become convinced that development was a do-it-yourself (DIY) process, one that had "to be open-ended rather than goal-oriented." Development should therefore be thought of as "an improvisational drift into unprecedented kinds of work" that would ultimately conclude by "drifting into improvised solutions."[31]

Such an approach moved beyond the quantifiable, ostensibly objective "industrial strategies" that marked nation-driven economic growth plans, strategies that sought to meet "targets" using "resolute purpose" and "long-range planning." These plans seemingly drew from the realm in which the state could stress its continued relevance: the arena of war. These state-driven models of economic development did little more than "express a military kind of thinking," one that was predicated on "a conscious or unconscious assumption that economic life can be conquered, mobilized, bullied, as indeed it can be when it is directed toward warfare."[32]

Instead, Jacobs argued that we must come to see the city itself as more human and development itself as more humane: "The order at work is more like biological evolution whose purpose, if any, we cannot see unless we are satisfied to think its purpose is us." Yet again, creativity was at the heart of this evolutionary process. Correcting decades of incorrect planning "depends on fostering creativity in whatever forms it happens to appear in a given city at a given time." This idea of nurturing development led Jacobs to stress what would come to be known as *social capital*—a term that Robert Putnam and others came to see Jacobs as playing an integral role in creating.

Sociability and social relationships mattered as they stoked creativity and, in time, economic development.[33]

Beyond Jacobs: Other Early Influences on the Evolution of Placemaking

Throughout the 1960s and onward, Jacobs often cited the influence of the architect Kevin Lynch on her work—as did placemaking advocates who would come after her. Unlike Jacobs, Lynch was a trained architect, one whose style of prose did not always prove accessible to the readers who flocked to Jacobs's writings. Yet their points of view complemented each other well, as Lynch's *The Image of the City* (1960) urged readers to "consider the visual quality of the American city," with close attention to be paid to "the apparent clarity or 'legibility' of the cityscape." How to help urban residents "read" the city—in an attempt to make their environments more livable—thus became Lynch's chief goal.[34]

Like Jacobs, Lynch saw little use for large-scale, one-size-fits-all planning in such an endeavor. To Lynch, there were "dangers in a highly specialized visible form"; therefore, there was "a need for a certain plasticity" in the urban built environment. As he found: "If there is only one dominant path to a destination, a few sacred focal points, or an ironclad set of rigidly separated regions, then there is only one way to image the city without considerable strain. This one may suit neither the needs of all people, nor even the needs of one person as they vary from time to time. An unusual trip becomes awkward or dangerous; interpersonal relations may tend to compartment themselves; the scene becomes monotonous or restrictive."[35] As the city clearly becomes a place meant to inspire sociability, Lynch came to call for a sense of aesthetic playfulness to inform the urban design process, as there is "some value in mystification, labyrinth, or surprise in the environment." At the same time, however, a city should not be too playful; there had to be limits to the ways that it mystified users. Too many surprises would put urban residents eternally on guard, unable fully to explore the relationships that the city should be able both to create and to foster. A city, in other words, should not make itself too confusing. Or as Lynch concludes in *The Image of the City*: "There must be no danger in losing basic form or orientation, of never coming out."[36]

This commitment to a visually compelling yet spatially ordered urban environment would lead to what Lynch believed to be the ideal social environment. "Indeed," as he noted, "a distinctive and legible environment not only offers security but also heightens the potential depth and intensity of human experience." As it helped foster such human relationships, such an environment would come to create a concept of *place* (a word that Lynch relied on heavily throughout *The Image of the City*) that would move beyond the micro-level of individual relationship to something even bigger: the creation of true community. It was here that "the sense of community may be made flesh," with the city becoming a location where "the citizen can inform it with his meanings and connections." The end result of this conversation of sorts between the individual resident and the surrounding urban environment? The city became "a true place, remarkable and unmistakable."[37]

Within the realm of architecture, this relationship between the social and space reached its apex in the 1970s with Christopher Alexander, who began this work in the 1960s and remains a strong influence for the placemaking movement of the early twenty-first century. With his 1965 essay "A City Is Not a Tree," Alexander began to explore the differences between *natural* and *artificial* cities. Somewhat confusingly, he compared artificial cities to trees, meaning that "within this structure no piece of any unit is ever connected to other units, except through the medium of that unit as a whole." This led to systems of the city existing "in isolation, cut off from the other systems of the city." As Alexander noted: "The city is

not, cannot and must not be a tree." Instead, it should strive to be a "semilattice," a structure marked by overlapping, interactive components that allow for multiple connections to develop over time. In such a structure, relationships matter more than anything else—and cities cannot develop without understanding their surrounding contexts.[38]

Here, one sees the roots of Alexander's concept of an architectural *pattern language*, a vocabulary that allowed for structures and spaces to exist—indeed, "speak"—as if they were human. Alexander would fully articulate this idea in his 1977 *A Pattern Language: Towns, Buildings, Construction*, alongside the thought that language helped us construct our identities through such places. One is struck by the ways Alexander connected the built environment to the construction of individual identity. At the root of this relationship was the realization that "people need an identifiable spatial unit to belong to."[39] However, the prevailing languages of twentieth-century architecture and planning served only to stifle individual growth. "The homogeneous and undifferentiated character of modern cities," Alexander concluded, "kills all variety of life styles and arrests the growth of individual character."[40]

Such a conclusion led Alexander to argue against the folly of overzealous urban zoning laws and for the embrace of mixed-use planning. Interestingly, he did not base his argument on the perceived economic value of such a shift in planning strategy. Instead, he saw mixed-use planning as a boon to the development of individual identity. Here was a sort of continuation of the personal crisis faced by Whyte's organization man. "The artificial separation of houses and work," as Alexander found, "creates intolerable rifts in people's inner lives."[41]

Yet Alexander was also intensely interested in creating spaces that helped cultivate the "outer lives" of city dwellers. This was also important for the overall health of both cities themselves and the residents that called them home. With this in mind, Alexander called for spaces that nurtured openness and sociability. One sees this in his discussion of the value of an urban promenade. Each group of urban residents, he felt, "needs a center for its public life: a place where you can go to see people, and to be seen."[42]

Alexander's most important legacy for the contemporary placemaking movement may be in the methods he prescribed to create such environments. To Alexander, such a process was "implemented best by piecemeal processes, where each project built or each planning decision made is sanctioned by the community according as it does or does not help to form certain large-scale patterns." To drive this point home, he insisted that such a process should not be expert driven—and it should not be top-down in its approach. "*We do not believe that these large patterns,*" he concluded, "*which give so much structure to a town or of a neighborhood, can be created by centralized authority, or by laws, or by master plans.*" Instead, they had to "emerge gradually and organically, almost of their own accord": This focus on the organic quality of space has become a central tenet of the contemporary placemaking movement, as has the idea that centralized authority has no place in the process. More than forty years after its publication, *A Pattern Language* remains a work frequently cited by contemporary placemakers.[43]

Prescribing Sociability: Public Space and Excess Individualism

The same year that *A Pattern Language* was published also saw the release of another work that would come to influence the development of placemaking: Richard Sennett's *The Fall of Public Man*. Like Alexander, Sennett was acutely interested in the relationships between public spaces and individual identity. And, much like both Whyte and Jacobs before him, he felt that we had come to privilege the private realm as crucial to these questions—at the expense of public, urban spaces. "We have tried," he wrote, "to make the fact of being in private, alone with ourselves and with family and intimate friends, an end itself." In such a society, "intimate feeling is an all-purpose standard of reality"—a reality that leads to overlooking public spaces. According to Sennett: "Intimate vision is induced in proportion as the public domain is abandoned as empty. On the most physical level, the environment prompts people to think of the public domain as meaningless. This is the organization of space in cities.… [T]he public space is an area to move through, not be in."[44]

Like Jacobs, Sennett saw the roots of these changes in the post–World War II United States, but, while Jacobs located the engine of change in government policy, Sennett saw the changing mores of the postwar generation as crucial. The generation that came of age in the 1960s "turned inward as it . . . liberated itself from sexual constraints." This move shifted attention to the value of the private sphere, leading to a simultaneous development in which "most of the physical destruction of the public domain has occurred." In such a shifting environment: "Community has become both emotional withdrawal from society and a territorial barricade within the city." This "warfare" replaced earlier physical manifestations of the public/private relationship. "This new geography," Sennett concluded, "is communal versus urban; the territory of warm feeling versus the territory of impersonal blankness."[45]

Again like Jacobs, Sennett saw the answer to this crisis in reinvigorating a certain type of urban sociability, one that saw great value in the impersonal relationships that can flourish in the city. People must once again find "great pleasure in that world of strangers" that took form in the "community relations of strangers." As he concluded: "The extent to which people can learn to pursue aggressively their interests in society is the extent to which they learn to act impersonally. The city ought to be the teacher of that action, the forum in which it becomes meaningful to join with other persons without the compulsion to know them as persons." The twenty-first-century placemaking guru Richard Florida would run with this idea some twenty-five years later.[46]

The next decade saw attempts to address these realms, with the work of people like the sociologist Robert Bellah and his influential *Habits of the Heart: Individualism and Commitment in American Life* (1985) coming to drive the discourse. The work began with the admission that the "central problem" of the book "concerns the American individualism that Tocqueville described with a mixture of admiration and anxiety." While not discounting the importance of such individualism, Bellah and his coauthors were "concerned that this individualism may have grown cancerous—that it may be destroying those social integuments that Tocqueville saw as moderating its more destructive potentialities." Something had to be done, in other words, to revive such moderating forces. Otherwise, Bellah and his coauthors concluded ominously, "the survival of freedom itself" was not a certainty for the United States.[47]

Like Sennett and many others who came before them, Bellah and his coauthors offered a lament for the loss of shared, public spaces, or what they broadly referred to as *community*. Such a sense of community was done away with through "the division of life into a number of separate functional sectors." Such division was rooted in economic imperatives as increased industrialization lead to working life becoming "more specialized and its organization tighter." Talking about the bureaucratic world that Whyte saw as on the rise in the 1950s, Bellah and his coauthors found that, by the 1980s, such "division suited the needs of the bureaucratic industrial corporations that provided the model for our preferred means of organizing society."[48]

At the same time, *Habits of the Heart* also put some blame on the long-term impact of urbanization in the United States: "The associational life of the modern [i.e., late twentieth-century] metropolis does not generate the kinds of second languages of social responsibility and practices of commitment to the public good that we saw in the associational life of the 'strong and independent township.' The metropolitan world is one in which the demands of work, family, and community are sharply

segmented and often contradictory, a world of diverse, often hostile groups, interdependent in ways too complex for any individual to comprehend."[49] Here, the impersonal metropolis lauded by Sennett was recast as an environment antithetical to the creation and maintenance of community. Which school of thought is best suited for placemaking efforts? As noted above, the contemporary urban theorist Richard Florida—himself self-consciously drawing from the earlier work of Jane Jacobs—has found great value in the loose ties of urban life. Other voices in the discussion of twenty-first-century placemaking efforts have, as we will see in chapter 3, argued for the need for a deeper sense of community, one along the lines suggested by Bellah and his coauthors. Such differences highlight the confusion present in efforts to define what placemaking truly is, making it all the more difficult to assess what these ideas should ultimately mean—and how they should operate.

But these dueling approaches to community building were not what was at the heart of the work of Bellah and his coauthors. Instead, they were interested in detailing the feeling of disconnection that rampant individualism had created within many Americans. To deal with such anxieties, such individuals—according to Bellah and his coauthors—embraced the cultures of the manager and the therapist. As work became increasingly central to individual identity, it is perhaps easy to see why the manager culture was adopted so readily. Yet such a move had real psychological impact, hence the rise of the therapist culture. But this culture also focused exclusively on the realm of the individual. At the same time, social concerns had come to trump economic concerns. "Life's joys and deeper meanings, and its difficulties also, are less often attributed to material conditions," Bellah and his coauthors noted. "Now the 'interpersonal' seems to be the key to much of life." This leads to stress on personal growth but as a purely private matter as "the very language of therapeutic relationship seems to undercut the possibility of anything other than self-interested relationships."[50]

Ultimately, Bellah and his coauthors proved better at diagnosing a potential problem than at offering any sort of tangible cure. They saw American culture, by the mid-1980s, as at "a profound impasse": "Modern individualism seems to be producing a way of life that is neither individually nor socially viable." Yet there seemed no clear-cut ways to revive a sense of community that could temper such individualistic impulses. Was it possible, in other words, to revive older civic values and traditions in ways that would speak to the denizens of late twentieth-century America? And, if so, what would such an effort come to look like?

Bellah and his coauthors were not the only ones discussing the costs of perceived out-of-control individualism. And here the focus became spatial, paying closer attention to surveillance and spaces of the city. One of the most influential of such works was in the world of policing. In 1982, George L. Kelling and James Q. Wilson published their foundational "Broken Windows: The Police and Neighborhood Safety." By the early 1980s, there was a perceived growing threat of broad crime and violence in American cities. But there was also something more specific and definitely less dangerous: "the fear of being bothered by disorderly people." Such "disreputable or obstreperous or unpredictable people" as "panhandlers, drunks, addicts, rowdy teenagers, prostitutes, loiterers, the mentally disturbed," who were not exactly real criminals, affected average citizens' perception of the city or how they saw the urban environment around them. Much like Kevin Lynch some twenty years earlier, Wilson and Kelling understood that how a resident saw the city affected their level of comfort with their urban surroundings. And, again echoing Lynch, they believed that too much surprise meant that disorder could prevail. Such perceptions came to flow from "a sense that the street is disorderly, a source of distasteful, worrisome encounters." This led them to see that "disorder and crime are usually inextricably linked, in a kind of developmental sequence."[51]

Yet the way to counter such damaging individual behavior was through a reengagement with community. In a passage that sounded like it could have come straight from the pages of Jane Jacobs's *Death and Life of Great American Cities*, Kelling and Wilson argued that the "substantive problem" was: "How can the police strengthen the informal social-control mechanisms of natural communities in order to minimize fear in public places?" The answer was to move away from the individual and focus more on the communal. Yet Kelling and Wilson were anything but social progressives. In fact, they argued that the "rights revolution" of the 1960s had placed too much attention on the rights of the person; lost was any sense of the rights of the broader community. Here, the semi-communitarian impulse was infused with a healthy dose of neoconservatism. To create safe neighborhoods,

we must return to our long-abandoned view that the police ought to protect communities as well as individuals. Our crime statistics and victimization surveys measure individual losses, but they do not measure communal losses. Just as physicians now recognize the importance of fostering health rather than simply treating illness, so the police—and the rest of us—ought to recognize the importance of maintaining, intact, communities without broken windows.[52]

Cities across the United States—including New York, Boston, and Baltimore—came to embrace broken-windows policing techniques as laid out by Kelling and Wilson. We will return to this relationship between policing, public space, and placemaking in chapter 2, but for now it is worth noting that such tactics further suggested that, by the 1980s, there was a strong belief that they overall condition of urban physical spaces matters greatly to their general health and that the environment created by such conditions played a large role in both individual and community behavior. The city of the 1980s may have looked bad, but it could be reformed—and transformed.

Reenter Whyte: Rediscovering the Center

By the late 1980s, people like William H. Whyte were putting on their metaphoric scrubs and attempting to find clear strategies for reforming the city. In 1988, Whyte published *City: Rediscovering the Center*, the culmination of his Street Life Project, which he had begun in earnest in 1970. As he explained in *The Social Life of Small Urban Spaces* (1980)—which he termed the *prebook* for what would become *City*—that project began by simply "studying how people use plazas." Chapters in *City* with such titles as "The Social Life of the Street" and "The Design of Spaces" give a hint of where Whyte was coming from—and why contemporary placemakers gravitate to his work.[53]

City: Rediscovering the Center was appropriately titled. In it, Whyte made the case for why it was time to move beyond a model fueled by suburban expansion and rediscover the benefits of the central city. In New York in the 1980s, this was not necessarily an easy argument to make. Whyte conceded that, throughout much of the post–World War II period, "What has been taking place is brutal simplification. The city has been losing those functions for which it is no longer competitive. Manufacturing has moved toward the periphery; the back offices are on the way. The computers are already there." The city of the twentieth century had been planned for an economy that was no more. Faced with such a changing economic forecast, the city now must, according to Whyte, reassert its most "ancient" function: serving as "a place where people come together, face-to-face." What did this mean? "More than ever, the center [of the city] is the place for news and gossip, for the creation of ideas, for marketing them and swiping them, for hatching deals, for starting parades.... This is the engine, the city's true export." And that meant creating places that worked to create sociability. "A good new space," Whyte clearly stated, "builds a new constituency." More specifically, he saw the value in the addition of outdoor urban cafés. "If you want to seed

a place with activity," he noted, "the first thing to do is put out food." Food can be a part of what Whyte called *triangulation*, the "process by which some external stimulus provides a linkage between people and prompts strangers to talk to strangers as if they knew each other." To help with such a process, he also often called for more open urban spaces; walls, shrubbery, and other items meant to demarcate space—and make it safe—often served to make such sites inaccessible and hostile.[54]

Yet Whyte was also clear that such design moves helped with urban strategies of surveillance—and drew limits on who and who could not claim ownership of such places. In a chapter from *City* tellingly titled "The Undesirables," he proceeded from the admission: "The biggest single obstacle to the provision of better spaces is the undesirables problem." Here, he was not talking solely about criminal actors like drug dealers, robbers, and petty street bullies; he was also referencing "the winos, derelicts, men who drink out of half-pint bottles in paper bags." As they had for Wilson and Kelling, such symbols of disorder hindered sociability in places that could come to foster such important social relationships.[55]

How to handle such a problem? Since "good places are largely self-policing," Whyte found that "the best way to handle the problem of undesirables is to make a place attractive to everyone else." His work with PPS—an advocacy group started by "designers and social scientists" in 1975 (according to a prospectus for PPS, "a time when values [were] changing, placing a greater emphasis on the public environment")—provided an apparatus through which to test such ideas. The group saw itself as up to the task; with its mission of "helping bring new life" to urban neighborhoods, PPS put into practice Whyte's prescriptions for creating usable urban places. By the late 1970s, it had itself become a prominent actor in urban planning in a host of American cities and, more so than Whyte individually, had proved influential in both private- and public-sector discussions of the viability of urban public

The old city of Lille, France, at night, the kind of urban fabric celebrated by William H. Whyte and others.

spaces. While Whyte would die in 1999, the ideas that he articulated—and that PPS amplified—would only grow in influence during the rise of creative placemaking in the early twenty-first century.[56]

Like others before it, PPS placed an emphasis on the neighborhood as the central organizational unit in city planning. According to its 1979 "Proposal to Provide Public Space Improvement Assistance to Neighborhoods in New York City," its work proceeded from the belief that "one of the great strengths of New York City is its neighborhoods." Yet, following the fiscal crises that cities like New York had faced during the 1970s, neighborhood-centered public spaces had fallen into various states of disrepair. "If public spaces are to make an important contribution to the vitality, stability, and security of neighborhoods," PPS found, "then fundamental changes in use, safety, function, and control of these spaces must be made."[57]

One such fundamental change was a rethinking of the size of viable public spaces. Echoing the "small is beautiful" concept as articulated by the contemporary economist E. F. Schumacher, PPS came to advocate for smaller, nontraditional urban spaces. For example, by 1977, it saw great benefit in the *vest-pocket park*, which it defined as "a small open space, left in a block between buildings, which has been redesigned for use as a park"—like Greenacre Park, located in midtown Manhattan just east of Third Avenue. By the late 1970s, it was working on developing such spaces in cities such as New York, Cleveland, and Seattle. First and foremost, PPS saw such parks as social spaces, capable of meeting the social needs of a variety of city dwellers. Yet equally as important was that such spaces could do this work without straining city coffers. These parks, as PPS concluded, were "important means for providing recreational opportunities without the large capital expenditure required for the establishment of the more traditional 'central park.'"[58]

These spaces did more than save cities money; they began to serve as engines of economic development as well. Here, PPS was ahead of its time, as is evident in the

fact that contemporary funders often saw little financial value to the work that the group was doing. Such a conclusion was evident in a 1979 letter from the Rockefeller Foundation rejecting a funding request. Writing on behalf of the foundation, Marilyn W. Levy found: "While reshaping the quality of the urban environment through work projects with local community boards is vital to the quality of city life, the current grant directions relating to strengthening the city's economic base are being more narrowly defined." Yet to PPS founder and president Fred Kent, such a limited understanding of the relationship between sociability and financial growth overlooked what the group was trying to do. "I am sure that you are aware of the many studies that show that jobs follow people to environments with vitality and a sense of community," Kent wrote in response. "Since our program helps to make a community more livable, we feel our efforts contribute greatly towards the goal of economic development." By the early twenty-first century, Kent's words would become the mission statement of creative placemaking.[59]

This does not mean, however, that PPS did not try to stress the connections between economic development, sociability, and livable public spaces to potential clients. By the early 1980s, it was developing strategies to put in place such a philosophy, as evident in 1983's "Downtown Management: A New Role for the Private Sector." In this document, the group saw its primary goal as organizing the private sector—"large corporations, small businesses, merchants, and property owners"—and then "providing a management structure so that they can make downtown improvements and work more effectively with the public sector." To PPS leaders, such public/private partnerships held the key to creating vibrant urban public spaces. "When private and public efforts are combined," the report concluded, "their contributions complement each other and produce dramatic results."[60]

Within such partnerships, PPS saw itself as serving as a liaison between the two sectors. It would work with municipalities on "making design improvements" in order to create spaces where people would want to be. It would then coordinate with cities on "developing art and cultural resources" as mechanisms through which to draw people to such spaces. Finally, it would partner with the private sector on "creating a marketplace" to bring economic development to these newly activated public sites. The role of the private sector could not be overstated. "Creating a place where people will want to be," PPS found, "requires a focus on retailing, promotions, advertising, the development of markets, and entertainment."[61]

Such strategies were fleshed out further in a February 1980 report on the PPS Public Space Assistance Program. The goal of this program was to create partnerships between New York City and private developers to assist in the designing and planning of usable public spaces. On the one hand, these spaces had to be developed as sites of sociability where "a vital human dimension and quality" were explicitly on display. On the other hand, there was a stated belief that crucial to developing such sociability was increased economic activity. How best to bring people together? By developing "appropriate commercial activities to help strengthen the economy of the area."[62]

This emphasis on the relationship between commerce and viable public spaces also tied back to Whyte's belief that such spaces had to rely, in part, on self-policing. The presence of economic exchange would, as PPS explained in another document, be vital in "encouraging the presence of large numbers of the 'right' people" in such spaces. "Good design," coupled with economic activity (e.g., the sale of "good food" by nearby vendors), "would help to attract the people in whose company the 'undesirables' would feel uncomfortable." Yet self-policing could go only so far. Such spaces still needed "good maintenance," which often meant a police presence and, in some cases, private security forces.[63]

The PPS approach to cultivating a certain understanding of urban public spaces can perhaps best be seen in New York City's Bryant Park. Throughout the 1980s, the group highlighted its involvement in the park as a means of illustrating the benefits of its methodology. Even into the twenty-first century, Bryant Park remains, for PPS and other placemaking advocates, a model of a certain brand of placemaking done right.[64]

To Whyte, the initial goal of the park, which opened to the public in 1934, was that the space be "a refuge from the city, free from the hustle and bustle of pedestrians." Yet, by the late 1970s, the site had become underutilized, "except by undesirables." Drug dealers and other unsavory characters were taking advantage of the dense shrubbery and lack of entry points into the park to conduct their illegal transactions.

To deal with this, Whyte and PPS pulled together a coalition of civic groups and nearby corporations to reclaim the park. Writing in 1988, Whyte found: "The key aim is to open the park up to the street. Several new entrances are to be added; the labyrinthine internal layout is to be simplified to encourage pedestrian flows. The shrubbery has already gone." Or as a PPS publication from the twenty-first century discussed this evolution: "We found that what drug dealers and homeless people liked about Bryant Park was exactly what most other people didn't like." This included narrow park entrances and shrubbery that hid what was going on, "which was why homeless people found it a good place for sleeping." Looking back at the efforts to transform the space, PPS highlighted a legacy that included the widening of the park entrance, the removal of sight-blocking shrubbery, and the addition of stands selling food and other refreshments.[65]

By the mid-1990s, Bryant Park was run by the Bryant Park Restoration Corporation, whose largest corporate members included the cable television giant HBO and NYNEX, a regional telecommunications company. The park came to feature public art, including a statue by Alexander Calder, and HBO would show movies on an outdoor screen on warm summer nights. During the day, the park became a popular lunchtime spot. According to the sociologist Sharon Zukin: "It has become a place for people to be with others, to see others, a place of public sociability."[66]

Yet such usage came with increased surveillance. The Bryant Park Restoration Corporation hired its own security guards while pressing the New York Police Department to supply uniformed officers to patrol the park as well. The presence of such forces—along with the importance of their corporate employers to the planning of the site—suggested, in the eyes of Zukin, "an even more aggressive example of privatization" of public space than the city was used to seeing. For Zukin, there was danger in relying too much on the private sector for the development of such spaces. "It marks," she concluded, "the erosion of two basic principles: public stewardship and open access." Though many placemaking practitioners would overlook the nuances of Zukin's argument and focus intently on her conclusion that culture could drive urban economic growth, these issues—brought to the surface in places like Bryant Park in the 1980s—would not only come to inform placemaking efforts in the decades to follow; they would also complicate such efforts.[67]

Toward a Third Place: Ray Oldenburg and the "Great Good Place"

In many ways, the sociologist Ray Oldenburg's *The Great Good Place: Cafés, Coffee Shops, Community Centers, Beauty Parlors, General Stores, Bars, Hangouts, and How They Get You Through the Day*, first published in 1989, synthesized the ideas about public space that Whyte and others had been articulating for the previous thirty years. Clearly inspired by the work of Whyte, Oldenburg made the point that such spaces did not have to be as grand as Bryant Park to be successful. His chief contribution was, thus, to insist on the democratization of placemaking. Much like Jacobs's *The Death and Life of Great American Cities*, Oldenburg's book has become a must-read for the contemporary placemaking movement. Close to thirty years after its original publication, it continues to find a place in the bibliographies of all types of reports dealing with placemaking. And it is not hard to see why.

In clear, straightforward prose, Oldenburg managed to synthesize almost the entirety of the postwar literature on the power of place.

Like so many others before him, Oldenburg saw the relationship between individual and community as out of whack by the end of the twentieth century—particularly in the nation's cities. "The course of urban development in America," he noted, "is pushing the individual toward that line separating proud independence from pitiable isolation, for it affords insufficient opportunity and encouragement to voluntary human contact."[68]

Not that the suburb was the answer. In fact, it was the root of the problem. Such places offered safety, but they did not offer a sense of place: "Houses alone do not a community make, and the typical subdivision proved hostile to the emergence of any structure of space utilization beyond the uniform houses and streets that characterized it." Echoing William H. Whyte some thirty years earlier, Oldenburg reported that such spaces led to isolated men, unfulfilled housewives (the setting of Betty Friedan's *Feminine Mystique* [1963]), and troubled children. Without any sort of community or the "informal public life" that supports community, "people's expectations toward work and family life have escalated beyond the capacity of those institutions to meet them": "Domestic and work relationships are pressed to supply all that is wanting and much that is missing in the constricted life-styles of those without community." Such realities led to a society replete with "disorganization and deterioration" as the middle-class family of the late twentieth century had come to resemble "the low-income family of the 1960s": "The United States now leads the world in the rate of divorce among its population. Fatherless children comprise the fastest-growing segment of the infant population."[69]

Such stress did more than damage US families; it also damaged the American economy. Oldenburg cited estimates that the nation's industries lost from $50 to $75 billion annually because of absenteeism, company-paid medical expenses, and lost productivity. What was happening? "In the absence of an informal public life," Oldenburg concluded, "Americans are denied those means of relieving stress that serve other cultures so effectively."[70]

This is where the "third place," "the people's own remedy for stress, loneliness, and alienation," proved so important. This space, first and foremost, provided a vital type of neutral ground on which the informal, public relationships that are necessary to the health of the city—and Oldenburg references both Jacobs and Sennett in his discussion of the importance of such spaces—can take hold. Such neutral ground throughout cities and neighborhoods "offer the rich and varied association that is [cities'] promise and their potential." And what makes such places neutral? They are places "where individuals may come and go as they please, in which none are required to play host, and in which all feel at home and comfortable."[71]

Owing to such characteristics, the third place also served as a leveler. As such, it was open to the general public and did not set any sort of formal criteria for access. This leveling "sets the stage for the cardinal and sustaining activity of third places everywhere. That activity is conversation." Once again, sociability took center stage as the third place stood ready "to serve people's needs for sociability and relaxation in the intervals before, between, and after their mandatory appearances elsewhere." It was thus a sort of sphere separate from the worlds of home, work, and school.[72]

In terms of its physical appearance, the average third place was "typically plain," though with a mood that was "playful." Drawing from Jane Jacobs, Oldenburg found that third places were "most likely to be old structures ... frequently located along the older streets of American cities, in the neighborhoods or quarters not yet invaded by urban renewers." And—as the cumbersome subtitle of Oldenburg's book suggested—they were often spaces of consumption. Oldenburg spent a great deal of time discussing the third-space potential of the neighborhood tavern, where a "talking/drinking synergism" created a remarkably social space. These spaces ultimately provided such personal benefits as "novelty (which is characteristically in short supply in industrialized, urbanized, and bureaucratic societies)," perspective, and "spiritual tonic."[73]

On the one hand, these third spaces allowed for the personality of the individual to be fully expressed. They were "real places," as opposed to the "nonplaces" that marred the late twentieth-century urban landscape. "In real places," according to Oldenburg, "the human being is a person. He or she is an individual, unique and possessing a character." In nonplaces, however, "individuality disappears." Yet, by providing the room for individuality to flourish, third spaces thereby simultaneously assisted in the valuable process of community building: "Whatever improves social creatures improves their relations with others. What the third place contributed to the whole person may be counted a boon to all." Here, yet again, the symbiotic relationship between space, individual, and community was on display. The act of association created "good human relationships"—relationships that "industrial society," with its "high degree of specialization," had either overlooked or actively subverted. Within such communal spaces, "the fragmented world becomes more whole and the broader contact with life, thus gained, adds to one's wisdom and self-assurance." There was a reciprocity between community and individual here as well.[74]

Perhaps not surprisingly, Oldenburg saw political and economic repercussions emanating from the third place. Explicitly channeling Alexis de Tocqueville, Oldenburg saw third places becoming "parent to other forms of community affiliation and association that eventually coexist with them." Within the realm of economics, such spaces provide models of flexibility for a changing economic climate. Most of urban America provided a model of what *not* to do: "Even as our corporations now realize that their futures are jeopardized by imposing systems upon employees without their input, we continue to impose equally-flawed urban planning upon citizen-users as though their involvement in the process were not crucial to success." Third places provided an alternative to the creativity-crushing monotony of such increasingly economically unviable nonplaces.[75]

Ultimately, then, such problems were social in nature: the standard approach to urban planning actively discouraged people from coming together. Or as Oldenburg put it: "People and activities are compartmentalized and protected from the incursions and intrusions of that which is different from the singular function or particular segment or population for which the space was designated." Yet the type of social interactions that such places needed to create was far from deep. Invoking what he termed the *paradox of sociability*, Oldenburg found, similarly to Sennett's take on these arrangements: "One must have protection from those with whom one would enjoy sociable relations. One can't have them bursting into one's home or one's place of work or even have them around when one wishes they were not." The third place thus provided a mutually agreed-on space for social engagement.[76]

Finally, such spaces also helped order the urban environment. "The third place," Oldenburg noted, "where it remains, exercises its measure of control in community life." As with Jacobs, Kelling, Wilson, and Whyte, surveillance was a crucial part of this story. "One of the clear consequences of policies antagonistic to the social and recreational use of the urban public domain," Oldenburg continued, "is the loss of the monitoring function performed by responsible and law-abiding citizens. It is the ordinary citizen who tips the balance toward a safe public domain for the policing agencies of a free society are not adequate to the task. It is the substantial numbers of average people who provide the 'natural surveillance' necessary to the control of street life."[77]

What was the main culprit behind the lack of such third places? The problem did not seem to be money as since the end of World War II—the moment that Oldenburg saw public life in the United States as beginning to come under attack—Americans "have become more affluent." But that increased wealth has played a role in the diminishing of community as "money creates the illusion that one does not need people." Yet it was the state and misguided government policy that truly turned the tables. "Old neighborhoods and their cafés," according to Oldenburg, "taverns, and corner stores have fallen to urban renewal, freeway expansion, and planning that discounts the importance of congenial and unified residential areas." Outside central cities, new residential neighborhoods, built through "negative zoning codes," discouraged mixed-use planning. By the late 1980s, it was taken as gospel that the government did more harm than good when it came to taking the lead in the redevelopment of such urban places.[78]

Yet there was still need for some sort of policy, some way to develop these places more formally. For too long, Americans simply assumed that "places for connecting and associating would somehow naturally be there." This ran counter to the practice of a number of European countries that, according to Oldenburg, "took care to space enough *bier gartens* or bistros to gel their urban localities into a collective life." And, while Oldenburg himself did not make any policy recommendations, this question of the role of policy would come to inform placemaking efforts well into the twenty-first century.[79]

Oldenburg concluded his work with the belief that such efforts must be grounded in the spatial. Writing as the digital revolution was beginning to unfold, he saw little hope for the communal third place in any sort of "network" or "liberated community." Such places, he found, were "not defined in terms of location but in terms of the accumulated associations of a single individual. One's friends, acquaintances, and contacts, however scattered, constitute his or her network." Such a take on community overlooked the fact that "community is a collective reality that does not depend upon the inclusion or exclusion or any given individual"; it should not be defined primarily as a "personal phenomenon." All this conspired to create the belief that "incessant and excessive promotion of the individual and the idea that the good life is an individual accomplishment." At the same time,

Magers and Quinn bookstore, an Oldenburg-like third space in Minneapolis.

any sort of rootless community simultaneously "discourages collective effort, discounts collective effort, and obscures the fact that many good and necessary things can only result from collective effort." Going into the 1990s, Oldenburg concluded, physical spaces would continue to matter. The digital world might continue to grow, but people—particularly those in cities—would continue to crave social interaction.[80]

Perhaps more importantly, the concepts that Oldenburg was in many ways introducing into his work here—social networks and social capital—would enter into both academic and popular discourse in the following decade.

Moreover, the 1990s would see the field of creative placemaking emerge as a distinct practice in and of itself, one that would embrace these concepts—as well as Oldenburg's understanding of third places—in the effort to revitalize America's urban centers. At the same time, it is clear that Oldenburg's thoughts on the power of Great Good Places were informed by such individuals and organizations as William H. Whyte, Jane Jacobs, Richard Sennett, and PPS (whose website would come to feature a glowing feature on the influence of Oldenburg's ideas on its own work); he selectively cited and synthesized the work of all these actors at various points throughout *The Great Good Place*. As the field of creative place-making matured in the 1990s, theorists and practitioners alike would continue to practice such a strategy. In fact, throughout the 1990s, these ideas often came together to create a recipe for usable public space: add a little Jacobs, stir in a touch of Whyte, and mix in a hint of PPS. History, in other words, mattered—but it was often a very simplistic understanding of these histories that carried the day.

Such intellectual lineages also overlooked myriad histories that could have come to inform the health and functionality of urban spaces. None of these actors—or the organizations they influenced or even came to represent—really challenged the broader system of US capitalism in which their subject matter existed. For example, prominent activists like Saul Alinsky had no place in such discussions (despite his friendship with Jane Jacobs), nor did organizations that drew direct inspiration from his work, like the Chicago-based Woodlawn Association. And, if such mainstream works on urban place had little time for class distinctions, they had even less to say on the topic of race. The physical, often urban nature of the broad twentieth-century civil rights movement—from such place-based organizing strategies as the "Don't Buy Where You Can't Work" campaign (which led to the hiring of African American workers in such cities as Chicago and New York during the 1930s) to Martin Luther King Jr.'s articulations of a "Beloved Community" (one that strove for, in King's words, "genuine intergroup and interpersonal living")—did not influence such discussions. This was the case even when, by the late 1960s, African American architects like J. Max Bond Jr. and organizations like Architects' Renewal Committee in Harlem came to see, in the words of the architectural historian Brian D. Goldstein, urban space serving as "the medium through which black power adherents expressed their vision of the alternative future that would follow from racial self-determination," leading to a "more humane city that valued local decision making, existing inhabitants, and their vibrant neighborhoods and everyday lives."[81]

To many observers of the late twentieth-century city, the events that gripped a variety of American cities in the late 1980s and 1990s made such histories obsolete while simultaneously making the case that something new had to be tried. From the 1989 New York City Central Park wilding incident—which highlighted for many the savagery of a rising crime rate and the fact that there had been no eyes on the street to see what happened to the attacked jogger and to help her (some even compared the attack on the jogger to the infamous 1964 murder of Kitty Genovese)—to the violence that marked a number of midwestern cities by the early 1990s, there was a sense that the social bonds that held American cities together were seriously fraying. To many observers, the anonymity that marked the late twentieth-century city had created spaces where people no longer knew their neighbors. Such spaces needed to be reclaimed, and that could be done by encouraging sociability, by strengthening those weakened urban social bonds. First and foremost, such a development, as emphasized by Wilson and Kelling (by way of Jane Jacobs) and Oldenburg, would create a basic sense of safety in America's urban centers. Once safe, such communities could then be revitalized—with spaces of sociability and social engagement at their core. As the 1980s became the 1990s, advocates of the newly emerging field of creative placemaking would come to stress these relationships as initially explored and articulated by such influential theorists.[82]

But what did community revitalization via placemaking ultimately mean? Many of the cities suffering through a surge in violence by the early 1990s also endured great financial hardship during the era's recession. To a growing number of observers, the two were connected. As cities across the country struggled to make the transition from a manufacturing-based economy to a knowledge- and service-based economy, those hit hardest by deindustrialization—among them, African Americans socially, economically, and spatially segregated in central cities—came to engage in illicit economic activity, often by entering the violent drug trade. Yet many cities and their neighborhoods—even those untouched by drug violence—struggled to answer the question, How best to address the economic imperatives of the post-industrial city? As the 1990s progressed, the promise of sociability—so important to community safety—would come to be seen by creative placemakers as the corner-stone for urban economic redevelopment as well. This relationship would be strengthened as the economies of American cities improved throughout the 1990s and as these places became statistically safer. We now turn our attention to the role that creative placemaking itself—and the histories that informed this field—was perceived to play in such narratives of urban redemption.[83]

The Roaring '90s

(*Previous spread*) An April 2008 noon mass at St. Rochus Croatian
Catholic Church in Johnstown, Pennsylvania. Owing to declining
membership, the church would be closed and consolidated with five
other small churches the next year.

Despite Ray Oldenburg's assertion that economic affluence had increased in the United States since the end of World War II, his *The Great Good Place* had the misfortune of being published during a moment of intense global financial contraction. By the early 1990s, the United States found itself in an economic recession; unemployment reached nearly 8 percent in June 1992 as the recession came to claim over 1.6 million jobs across the country. Simultaneously, violent crime was reaching epidemic proportions nationwide. Aggravated assault had grown by 134 percent from 1972 to 1992, while handgun-related homicides had more than doubled between 1985 and 1990. In 1991 alone, Chicago saw 922 killings, which amounted to the highest per capita murder rate recorded in the city. The year prior, murders peaked in New York City, with 2,245 slayings reported.[1]

Yet, by 1993, the American economy appeared to be back on track. Indeed, by the end of the decade, many economists saw the domestic economy as flourishing. During the second term of President Bill Clinton (1996–2000), real economic growth averaged 4.5 percent per year, while unemployment fell to 4 percent, a level that—according to the economists Jeffrey Frankel and Peter Orszag—"had been specified by the Humphrey-Hawkins legislation three decades earlier to be the goal of national policy." By 2000, the poverty rate had fallen to approximately 11 percent, essentially a post–World War II low. In that same year, the US economic expansion surpassed in length the expansion of the 1960s, thus becoming the longest period of economic growth on record. Across the globe, economic globalization brought about surges in trade and capital mobility. Following the fall of the Soviet Union, there was a growing confidence that both democracy and free market capitalism were transforming the world.[2]

Such economic growth seemed to contribute to rising population growth in the United States. The 1990s saw a 13 percent increase in total population, a jump of 32.7 million people from 1990 to 2000. This surpassed the previous record growth of 28 million during the peak of the baby boom of the 1950s (though the growth rate of the 1990s was below the 18.4 percent growth rate of the 1950s).[3] And where was this growth happening? In American metropolitan areas anchored by large cities. Reporting in April 2001, the US Census Bureau found that metropolitan areas with populations of two to five million grew the fastest, up 20 percent. Observing such trends, the economists Edward L. Glaeser and Jesse Shapiro asked, in June 2001, "Is there a New Urbanism?" The numbers seemed to suggest that the answer was yes. In fact, eight of the ten largest US cities gained population in the 1990s, with only Detroit and Philadelphia declining in size. While a number of cities on this list, including Sun Belt cities like Phoenix and Houston, had been steadily growing throughout the post–World War II era, others, including New York City and Chicago, were adding people after shedding thousands of inhabitants throughout the 1970s and 1980s. By the end of the twentieth century, many of these cities had become what the sociologist Saskia Sassen has termed *global cities*. To Sassen, the spaces of such cities mattered as their landscape housed the firms integral to the infrastructure of the flow of capital and information or those things that fueled postindustrial economic growth.[4]

On the one hand, the relative stability of the 1990s, both domestically and internationally, created a moment during which the question that had vexed observers during the preceding decades—How best to build community?—could be more robustly answered. More specifically, as economic insecurity came to be perceived as less of a problem for more and more Americans, countless thinkers saw this as an opportune moment to rethink the relationships between the individual and the communal that had exasperated so many during the turbulent 1980s.

As economic concerns receded, culture was seen as central to this process of reconciliation. It would foster the human relationships central to the creation of true community. Yet these relationships needed a venue. Here, a newfound emphasis on the importance of space and, more particularly, urban space came to the fore. It was in such spaces that the sociability needed to cultivate productive human relationships could best be developed. Perhaps not surprisingly, then, it was during this era that both the power of place and the concept of placemaking came to be better defined.

Yet culture—and the placemaking efforts that soon sought to capture and represent it—also came to be seen as a driver of economic growth in many such cities, proving valuable in marketing the postindustrial city to a variety of actors. It could create both community and self-sufficiency in ways that moved the cost of certain urban amenities away from the municipality itself. And all such developments would provide avenues for economic growth, avenues desperately needed by numerous cities following decades of deindustrialization and capital mobility.

This chapter draws on the works that proved influential both at the time and during contemporary discussions on the practice of placemaking. Again, these ideas, rooted in the 1990s, remain relevant in the twenty-first century. Such movements as communitarianism and New Urbanism, along with such phenomena as social capital, community policing, culture-driven urban redevelopment, and placemaking itself, all came to be fleshed out more thoroughly during the 1990s. All these continued to inform placemaking in the early twenty-first century.

There is little doubt that the optimism associated with this historical moment has informed a hopeful definition of *placemaking*, one that sees great promise in both the American city and the US economy. At this time, the relationship between the city and culture came to be seen as a vital tool in renewing America's urban centers along with those often left behind by earlier renewal efforts. More specifically, creative placemaking—and such ideas as social capital that often undergirded creative-placemaking endeavors—was seen as a way to address the plight of people of color and others who suffered as a result of late twentieth-century deindustrialization. Here, the economic value of placemaking was privileged. As such, placemaking efforts came to allow culture, rather than such previous economic drivers as manufacturing, to become the focus of urban redevelopment strategies. Such strategies would only grow in popularity as the twentieth century drew to a close.

Yet the optimism of the moment missed one of the other features of Sassen's global city reorganization: many large American cities continued to lose population during the 1990s. Detroit's population, for example, went from 1.029 million in 1990 to 945,297 in 2000 (it would fall to 677,116 by 2015). Such population loss led to declining revenues for many cities, which, coupled with the fallout from the mass closures of industrial plants, led to continuing high rates of poverty for many urban communities, where continuing racial segregation only served to fuel such conditions further. Such realities would continue to haunt placemakers in the early twenty-first century.[5]

We're Not in This Alone: The 1990s and Communitarianism

By the late 1980s, a group of scholars, concerned by the trends documented by Robert Bellah and his colleagues, came to organize under the name *responsive communitarians*. Calling for a balance between individual liberty and social order, a group of fifteen ethicists, social scientists, and social philosophers met in Washington, DC, in 1990 and effectively launched the communitarian movement. The group, called together by the sociologist Amitai Etzioni and the political theorist William Galston, issued "The Responsive Communitarian Platform," the preamble of which stated: "Neither human existence nor individual liberty can be sustained for long outside the interdependent and overlapping communities to which all of us belong. Nor can any community long survive unless its members dedicate some of their attention, energy, and resources to shared projects. The exclusive pursuit of private interest erodes the network of social environments on which we all depend, and is destructive to our shared experiment in democratic self-government. For these reasons, we hold that the rights of individuals cannot long be preserved without a communitarian perspective."[6]

Even though such a perspective "recognizes both individual human dignity and the social dimension of human existence," there was a stated belief that individualism had been allowed to grow out of control.[7] What were therefore needed were ways to foster types of social engagement that could counter this damaging trend. By January 1991, the group was publishing a quarterly titled *The Responsive Community: Rights and Responsibilities*. In 1993, Etzioni formed the Communitarian Network, a nonprofit think tank dedicated to spreading the communitarian message.

Etzioni's 1994 *The Spirit of Community: The Reinvention of American Society* was perhaps the clearest articulation of the communitarian vision. Looking back at the 1990 conference, Etzioni saw a group of thinkers convinced that "free individuals require a community." Such communities, on the one hand, "back them up against encroachments by the state." Yet at the same time a strong community also "sustains morality by drawing on the gentle prodding of kin, friends, neighbors, and other community members, rather than building on government controls or fear of authorities."[8]

As Etzioni explained elsewhere, communitarians "argue that the preservation of the social bonds is essential for the flourishing of individuals and of societies." Moving away from the belief that "it requires the state to determine the good and then direct its laws toward promoting that good," he called for individuals to "pay special attention to social institutions, which form the moral infrastructure of society." He often privileged "families, schools, communities, and the community of communities," that is, those institutions that stressed sociability. The economic and political realms took a backseat to the world of the social.[9]

For Etzioni, then, the main goal of communitarianism was to create the "we-ness" that would balance individual freedom and communal support. This communal feeling would lead to "the new community." Such a community would be predicated on "a balance between liberty and social," and in it "individuals [would face] responsibilities for their families, communities, and societies—above and beyond the universal rights all individuals command, the focus of liberalism." To meet such responsibilities, communitarians needed to foster "social bonds" and grow "social institutions" (including schools, communities, and "the community of communities"): such things were the mechanisms through which we-ness could be achieved.[10]

Interestingly, Etzioni saw flashes of these new communities in urban centers across the United States. Without getting too specific, he found that "many cities have sustained (or reclaimed) some elements of community." Citing the work of James Q. Wilson, Herbert Gans, and Jim Sleeper, he argued that the "urban village" had often created bonds of reciprocity that lead to stronger communal ties. And these bonds seemingly transcend issues of economy. Drawing on the work of Diane Swanbrow, he noted that, even though real income for Americans rose from 1946 to 1978, there was no similar increase in happiness. "Social science findings," he concluded, "leave no doubt that genuine inner satisfaction cannot be attained that way."[11]

While looking at what helped foster community, Etzioni also privileged the impact of the American built environment. Writing on communitarian design, architecture, and planning, he noted: "To make our physical environment more community-friendly, our homes, places of work, streets, and public spaces—whole developments, suburbs, and even whole cities—need to be designed to enhance the Communitarian nexus. . . . An era dedicated to a return to we-ness would value and promote design that is pro-community." Unfortunately, he provided little detail on what these important spaces would actually look like. Instead, he recommended building on the earlier work of Jane Jacobs. He seconded her belief that buildings be integrated into the street and that "vest-pocket parks and gardens are more community-friendly then grand ones or those designed for decoration and effect."[12]

However, Etzioni exerted a great deal of effort laying out the values that such spaces would enhance. Once again, he came back to the need for spaces of enhanced sociability. At their core, these spaces of we-ness needed to "provide people shared space to mingle." Most importantly, they would provide the physical room for the creation of new personal relationships. "Some of my best friends," Etzioni wrote approvingly, "are those I met when I watched my sons' play in the sandboxes of Riverside Drive in New York City." Ultimately, these spaces worked because they "enhance rather than hinder the sociological mix that sustains a community." Such ideas would come to inform the placemaking movement of the early twenty-first century as references to Etzioni's writings made their way into a myriad of such efforts.[13]

Urban Culture and Reassessing the Power of Place

Others were coming to theorize space in similar ways at this historical moment. In Los Angeles, the urban historian Dolores Hayden's nonprofit organization the Power of Place sought to reclaim and reconnect people with their (often-forgotten) urban pasts. Hayden began to document this work in 1995's *The Power of Place: Urban Landscapes as Public History*, in which she argued: "Locating ethnic and women's history in urban space can contribute to what might be called a politics of place construction, redefining the mainstream experience, and making visible some of its forgotten parts." Seeking to expand what she termed the *politics of identity*, she came to see public space as a place to craft a more "inclusive sense of what it means to be an American."[14] For Hayden, the concept of place memory, which she borrowed from the philosopher

Edward S. Casey, allowed her to suggest a way for city dwellers fully to engage with the cultural landscape of a city by helping them connect with both the built and the natural environments. Place memory thus undergirded a certain type of urban placemaking that would "take on a special evocative role in helping to define a city's history" and, in the process, "begin to reconnect social memory on an urban scale."[15]

There is little doubt that Hayden did groundbreaking work, and her interventions within the built environment of Los Angeles provided a template for action for later placemaking efforts around the nation. Her effort to bring the story of the slave–turned–Los Angeles real estate entrepreneur Biddy Mason to life exemplified her approach. With her team, she worked to install an innovative public art memorial that captured the time line of Mason's life. Photographs, letters, drawings, maps, and even objects such as Mason's midwife's bag and a medicine bottle are embedded in a black concrete wall eighty-one feet long and eight feet high. Nearby is *Biddy Mason: House of the Open Hand*, a sepia-toned photomural done by the artist Betye Saar, a collaborator of Hayden's, that humanized Mason by showing her sitting with three women on the front porch of a nearby house.

Such efforts rescued certain histories from the constant redevelopment that marked Los Angeles during the 1980s and 1990s. Yet not all urban locations have proved as desirable for redevelopment as Los Angeles. Hayden argued that residents would better understand their cities' complex histories if they encountered presentations of those histories in the course of their daily lives. And what was the end result of such a process? "People invest places with social and cultural meaning, and urban landscape history can provide a framework for connecting those meanings into contemporary urban life." But such a conclusion raised the question, What about economic meaning, particularly in a city hit hard by deindustrialization?[16]

An epilogue titled "Los Angeles After April 29, 1992" somewhat unintentionally revealed the shortcomings of such an approach. Hayden's efforts during the 1980s did not really address the root causes of the 1992 uprising, and the best that Hayden could do in 1995 was note: "Choosing to engage the difficult memories, and the anger they generate, we can use the past to connect to a more livable future." But she was largely silent as to how to make such a connection and what such a connection would ultimately look like. Lacking such answers, she could conclude only: "While this book is about cultural possibilities, urban Americans need to find a new kind of political will to pursue them." While such a conclusion was ultimately less than satisfying, it did point out that the Los Angeles riots were a point of inflection for Hayden's thinking on urban space. This would come to be true for others thinking through similar issues throughout the 1990s.[17]

The possibility—indeed, power—of urban culture was explored in another influential book from 1995, Sharon Zukin's *The Culture of Cities*. Unlike Hayden's, Zukin's work very quickly dealt with the intense economic change that had been occurring in America's urban centers. Yet these momentous shifts were taken almost as a fait accompli, with little indication of just how wrenching—and life changing—they were for many US cities. As manufacturing industries continued to disappear from the urban landscape, taking with them the revenue streams that fueled local government, culture became "more and more the business of cities." And the growth of culture industries—including food, art, fashion, music, and tourism—was fueling "the city's symbolic economy, its visible ability to produce both symbols and space." This reordering did not necessarily challenge the primacy of capitalism; instead, it changed the way capitalism was lived by both consumers and employees alike in urban centers across the country. The implication here seemed to suggest that cities needed to keep up or get left behind.[18]

Within such a new urban world Zukin saw culture both as an economic base and as a means of framing space. Such developments could be positive as they opened up spaces previously closed to certain groups. The industrial era was, after all, one of intense segregation and economic inequality, and these conditions often rendered people of color invisible, lacking visual representation in such important arenas as American politics, economics, and culture. Yet, with its return to the city and emphasis on culture rather than manufacturing as the prime economic engine, the postindustrial era seemed to provide a space where those previously marginalized could be seen. There was an obvious financial element to such an evolution: the creation and growth of hip-hop, for example, offered a new product that could be marketed to music fans across the globe. But as Zukin put it: "Incorporating new images into visual representations of the city can [also] be democratic. It can integrate rather than segregate social and ethnic groups, and it can also help negotiate new group identities." To make this point, she illustrated how African American organizations in Pasadena, California, successfully demanded representation on the nine-person commission that managed the annual and highly visible Rose Parade. These organizations used culture rather than legal mechanisms to bring about integration. "By giving distinctive cultural groups access to the same public space," Zukin concluded, "they incorporate separate visual images and cultural practices into the same public culture."[19] By 2019, this display of public culture featured marching bands from historically Black colleges and universities, performances by African American musical acts like Kool and the Gang, and grand marshals such as Chaka Khan. In 2018, an African American man—Gerald Freeny—was named president of the Tournament of Roses, the body that organizes the Rose Parade.[20]

Elsewhere, Zukin saw the power of culture being harnessed to aid redevelopment in such postindustrial cities as Flint, Michigan, Camden, New Jersey, and North

Adams, Massachusetts. In North Adams, the Massachusetts Museum of Contemporary Art (MASS Moca) successfully took root in a formerly abandoned factory complex and, ultimately, opened to the public in 1999. Yet similar efforts in Flint (Autoworld) and Camden (the New Jersey State Aquarium) either failed spectacularly or were nowhere near as successful as expected. While Zukin suggested that race mattered, she did not pay close attention to how and why it was essential (an approach shared by contemporary placemakers). As she wrote: "Camden has an almost entirely black and Latino population; Flint also has a significant population of ethnic minorities. Aside from its mainly white population, what makes North Adams different?" That's a huge difference, but it also pointed to things that often follow whites: educational and cultural capital. Nearby Williams College helped MASS Moca get off the ground, as did such cultural institutions as the Guggenheim Museum. Neither Autoworld nor the New Jersey State Aquarium had such resources on which to draw. Not all urban cultures—and corresponding institutional networks—were the same.[21]

Because of such disparities, redevelopment occurred in fits and starts in a broader atmosphere marked by intellectual chaos and real-world design confusion. As Zukin concluded, such work was being done in the aftermath of "the death of urban planning," which had led to "unprecedented debates over what should be built (or unbuilt) and where." More specifically: "A deep chasm lies between the post-1970s appreciation of visual culture and the absence of a master vision to control the chaos of urban life." Here was both a crisis and an opening: "This gap offers opportunities for both access and exclusion, for both elitism and democratization." Writing in 1995, Zukin was not yet sure which side would win or even how this process would look as it played out. Ultimately, she ended her work on an ambivalent note: "Who can tell where all this leads?" Soon, others would attempt to answer this question as the act of placemaking came to be seen as a new way to harness culture to provide a new vision of ordered urban life.[22]

The Rebirth of Planning? The Professional and Placemaking

As Zukin noted, she was writing in the aftermath of a perceived demise of urban planning. Scholars have traced this demise back to the 1960s, when attacks on the field by the likes of Jane Jacobs and her allies brought it to its knees. Postwar urban planners had "abetted some of the most egregious acts of urban vandalism in American history"—from the wholesale destruction of New York City's Manhattantown project to the reality of such inhumane housing projects as St. Louis's Pruitt-Igoe—and the subsequent death of the field opened it up to myriad new actors and voices. Yet, according to historian Thomas J. Campanella, "Privileging the grassroots over plannerly authority and expertise meant a loss of professional agency."[23] By the beginning of the 1970s, professional planners and academic departments devoted to planning found that "the basic assumptions of the planning profession [were] being called into question."[24]

Things did not get much better for the field during the following decades. By the late 1980s, Michael P. Brooks observed, in an influential essay, that planning was suffering through a prolonged identity crisis, one that saw the field trying to address a host of concerns, including "the quest for increased political efficacy," "the quest for sanction and financial support from the federal government," "the quest for academic respectability," and "the quest for private-sector validation." To Brooks, the tensions between such concerns had to be ameliorated in order to salvage "the soul of the profession." Such tensions remained unresolved through the early twenty-first century, leading to what Thomas Campanella has termed a professional impotence among urban planners. What could be done to revive this flailing field, to make it relevant once again?[25]

For some, the answer to such questions was the emerging concept of placemaking. The year 1995 also saw the publication of the architecture professors Lynda H. Schneekloth and Robert G. Shibley's *Placemaking: The Art and Practice of Building Communities*, a work that seemed to find a way for planning to address all the concerns that Brooks had laid out in 1988. Similarly to Hayden's and Zukin's, Schneekloth and Shibley's concept of placemaking was "embedded in the concept of *place* and [grew] out of a critical practice of making space." While their concept was rooted in practice, they ultimately put forward a fairly theoretical idea of the concept rooted in social relationships as placemaking became "a fundamental human activity," "the way all of us as human beings transform the places in which we find ourselves into places in which we live." This could involve the mundane, like rearranging a home or an office, or the dramatic, such as tearing down structures and building new ones. Moreover, placemaking could be a long-term process or a onetime occurrence. It could be carried out with the support of government and other institutions, but it also could be "an act of defiance in the face of power." Ultimately, however, it was all about creating spaces of sociability. "In other words," as Schneekloth and Shibley concluded, "placemaking is not just about the relationship of people *to* their places; it also creates relationships *among* people in places."[26]

Such an idea was undoubtedly informed by the rise of postmodernism in such fields as architecture and urban planning at this time. As Schneekloth and Shibley found: "The social construction of knowledge and place removes them both from the private realm and relocates them in a public and relational practice, a dialogue." Within this dialogue, both community members and professional placemakers could "'legitimately' construct social and place-relevant knowledge to enable and empower communal action." Professionals—with their specific skill sets—had much to add to such a conversation, but the postmodern turn showcased the danger of any sort of imposition of their knowledge over a certain space. Instead, when done right, the practice of placemaking "engages the activities of situating and translating the knowledge of all participants within the dialogic space." When practiced through the act of creative placemaking, planning once again became empowering for the professional.[27]

Yet what was perhaps most noteworthy about their work was that Schneekloth and Shibley provided a new role for the professional planner alongside other "professional placemakers" such as architects, building tradespeople, facility managers, interior designers, engineers, and landscape architects. The work of such professionals was critical to placemaking. But such actors should not be the only ones involved, and they should not run the show as "the allocation of such work to a small body of professionals is fundamentally *disabling* to others." There had to be both the desire and the means to engage with the public at large, whose lives, ideas, and skills must be considered in all placemaking endeavors.[28]

What emerged in *Placemaking* was an attempt to bridge this divide between the professional and a variety of audiences, including community members. On the one hand, "the appropriation of placemaking activities by professionals denies a fundamental human expression." Yet on the other hand, Schneekloth and Shibley "recognize[d] that professional placemakers have something important to offer." Those with "expertise" could further this "universal activity" by "enabling and facilitating others in the various acts of placemaking," serving as a sort of conduit between the community and a body of ideas and resources. One party was not more important than the other; theirs was a relationship—one among people who usually do not work together—based on equality and reciprocity. For academics and other professional practitioners, such a conclusion proved revelatory. Once again, there seemed to be a place for them in discussions of urban planning and the urban built form.[29]

To drive home the point that such fledgling relationships could get things done, Schneekloth and Shibley documented their participation in the design and construction of a new church in Roanoke, an office redesign, and a planning effort called the Roanoke Neighborhood Partnership. Yet, rather than focus on the nuts and bolts of such placemaking work, they instead continually highlighted the ways that such work created a public space where dialogic partnership could occur. Placemaking thus "offers a unique public space in which to weave a web of relationships that interact to create a common world": "This is neither a space of 'public opinion' nor solely a collection of individual votes, but a sustained and shared public dialogue about who we are and where we want to live." According to such an assessment, placemaking was "not only about the physical making, remaking, and unmaking of the material world." Instead, it was about "'world making' in a much broader sense because the practice literally has the power to make worlds—families, communities, offices, churches, and so on." Once again, placemaking was vital because of the human relationships it could foster.[30]

Schneekloth and Shibley were well aware that people would not see such mundane acts as designing a church and working on a regional development plan as "world making." However, they "intentionally use[d] [those] words and their power to focus attention on the potential of critical placemaking to construct the world." What mattered was not necessarily the size or the scope of the project. Instead, it was that it allowed "for genuinely satisfying labor by professionals and people in places." Here, Schneekloth and Shibley essentially invented importance for placemaking efforts of all kinds, arguing that even small projects could be world changing. Such a belief— that size does not matter—came to inform placemaking efforts into the twenty-first century.[31]

Here Come the New Urbanists

The work of Schneekloth and Shibley suggested that the postmodern turn could be potentially liberating for those looking for an alternative to the orthodoxies of modern planning. Yet, for some practitioners, the types of planning and architecture that reigned supreme during the late twentieth century betrayed the promise that postmodernism had displayed on its arrival in the early 1970s. By the 1990s, a slew of "starchitects," including Michael Graves and Frank Gehry, had created a collection of highly stylized monuments to private wealth, both individual and corporate. The belief that postmodernism could foster a democratic dialogue to inform how public spaces could look and work was overshadowed by this turn of events as postmodernism became shorthand for little more than an aesthetic approach to architecture and planning.

Not all architects and planners, however, were happy with such a state of affairs. As postmodernism focused more intently on both individual talent and individual wealth, new approaches to these fields emerged to provide more specific blueprints for both communal design decisionmaking strategies and communal spaces. One group driving such discussions was the New Urbanists, a group that, since the early 1990s, has been a vital influence on the discourse surrounding contemporary placemaking. The New Urbanism story began in 1991, when the Local Government Commission, a nonprofit organization based in Sacramento, California, invited the architects Peter Calthorpe, Michael Corbett, Andres Duany, Elizabeth Moule, Elizabeth Plater-Zyberk, Stefanos Polyzoides, and Daniel Solomon to come up with a series of best principles for land-use planning. The result of this collective endeavor was the Ahwahnee Principles (a sample principle: "Public spaces should be designed to encourage the attention and presence of people at all hours of the day and night"), which the group then presented to over one hundred government officials in the fall of 1991, at the first Yosemite Conference for Local Elected Officials. Building off the excitement generated by such an effort, most of the group went on to form the Congress for the New Urbanism (CNU) in 1993.

In 1996, CNU members ratified the Charter of the New Urbanism. Through twenty-seven principles, the group laid out its commitment to providing an alternative to the suburban sprawl that they saw as threatening to take over the nation. At its core, the charter called for the restructuring of public policy and development practices to support a series of overarching goals: "Neighborhoods should be diverse in use and population; communities should be designed for the pedestrian and transit as well as the car; cities and towns should be shaped by physically defined

One of Michael Graves's much-lauded postmodern public buildings, the Denver Central Library.

and universally accessible public spaces and community institutions; urban places should be framed by architecture and landscape that celebrate local history, climate, ecology, and building practice."[32]

This attention to the primacy of neighborhood as an organizational unit was woven into much of the charter—as was a particular understanding of what these units should look like. A kind of stylistic conservatism was also woven through the document, emphasizing how new development "should respect historical patterns, precedents, and boundaries" and how "individual architectural projects should be seamlessly linked to their surroundings." The group claimed that these guidelines should "tran-

scend style," but in work that followed the emphasis on street-width-to-building-height ratios, tree canopies, and form-based codes resulted in an aesthetic yearning for the early twentieth-century American small town, with a touch of Jane Jacobs's Greenwich Village thrown in for good measure.[33]

Still, CNU was not simply about style. Planning was also central to its mission: "Neighborhoods should be compact, pedestrian-friendly, and mixed use."[34] In subsequent writings, key figures such as Duany and Plater-Zyberk noted: "There are two types of urbanism available: the neighborhood, which was the model in North America from the first settlements to the second World War; and suburban sprawl, which has been the model since then." The goal of New Urbanism was to lead the revival of this former model of American urbanism.[35]

Why revive the focus on neighborhoods? For Duany and Plater-Zyberk, such spaces best bred true community, a community that stressed—and provided the spaces for—increased sociability and the additional benefits, such as security, that came with such human interactions. As they noted: "By providing streets and squares of comfortable scale with defined spatial quality, neighbors, by walking, can come to know each other and to watch over their collective security."[36]

This notion of creating spaces that fostered community was further driven home in the CNU architect Philip Langdon's "Can Design Make Community?" of 1997. Langdon found that "community *layout*," rather than style of architecture, was crucial to the goals of New Urbanism. And, in such a community, "the individual buildings work together to form coherent public spaces, where people will see and talk with one another." In other words: "New Urbanists seek more than pleasant aesthetic effects—they aim to create a different, more neighborly social life within their communities."[37]

Interestingly, the creation of communal spaces would lead to individual growth. Langdon found that the "automobile-dependent suburbs" have overlooked the fact that "youngsters need a modulated introduction to the world beyond their block, so that they can cope with, and learn to thrive in, a country that has never been, and never will be, entirely safe or homogeneous." By countering the suburban push to "withdraw its children from society's difficulties," New Urbanism aimed to foster more mature, more capable young people. As design essentially forced young people to engage with the world around them, they grew individually.[38]

What apparently gets in the way of such growth? Government. As Langdon concluded: "The inflexible hand of zoning should be loosened so that mixtures of housing—gradations of different kinds, sizes, and prices—can be included in a community, and so that small stores, ca-

fes, and other hallmarks of a sociable neighborhood can legally be built within walking (or biking) distance of people's lives."[39]

Plater-Zyberk also noted that, in the early days of the movement, "policy was precluding certain efforts." This led her to dig deep into the nitty-gritty of zoning and building codes. Such a focus moved her away from any sort of intellectual discussion of the theory of place. Instead: "We were more interested in technique." She saw New Urbanism as attempting to create a "certain ambience … an aesthetic response to the divisive suburb." She therefore saw it initially as an "aesthetically-based community." But she soon also saw that it was also "a social thing," something that could expedite "community building." And there was a focus on building: the early New Urbanists had little interest in simply mimicking the work of the preservationists. Commenting on the preservation movement, Plater-Zyberk expressed admiration but also a desire to do more. "They were the first ones to stand up and say 'no,'" she commented. "We were more interested in how do we get to 'yes'?" Most importantly, she saw the process as democratic: "We championed public participation." This was a break from earlier planning processes, where, with the support of political leaders, urban planners made planning decisions far removed from any sort of real public discussion. This commitment to the perceived value of public participation continues to inform placemaking efforts in the twenty-first century.[40]

And all this led to a new engagement with the city—and new ways to think about it, past a superficial postmodernism that was still very much in vogue. For the more academically inclined, New Urbanism provided a vehicle through which to think about the city and its past in nonironic ways. As Plater-Zyberk noted, the work of the New Wave of European architecture and the Institute for Architecture and Urban Studies (which brought the European architects to the United States in the late 1970s) made her think of "the city as the venue of architecture, not just the individual building." At the same time, the postmodern turn led to her own personal discovery of history. Yet she wanted to do more than what she saw as the shallow examination of history offered by leading postmodernists like Robert Venturi; she undertook what she called a "very conscious examination of history." Through this process, she looked back into the 1910s and 1920s, particularly in Europe, which brought a "Garden City influence" to her thoughts on urban design. New Urbanism provided the opportunity for this new assemblage.[41]

Yet practitioners of placemaking have not necessarily been drawn to the nuanced understanding of place offered by the likes of Langdon and Plater-Zyberk. Instead, they are drawn to the more straightforward work of people like James Howard Kunstler. Kunstler's *The Geography of Nowhere: The Rise and Decline of America's Man-Made Landscape* (1993), cited often by contemporary placemakers, is a polemic that perhaps best captures not only the ideas but also the passion that motivated New Urbanists. In this work, Kunstler looked out at a late twentieth-century "landscape of scary places, the geography of nowhere," that had "simply ceased to be a credible human habitat." Resurrecting a 1960s-era bogeyman, he blamed such conditions—particularly as they existed in American cities—on a government-sponsored modernism that was "a response to the rise of industrial manufacturing as man's chief economic activity."[42]

As urban renewal took its toll on the urban landscape, more and more individuals sought refuge in the nation's burgeoning suburbs. But, for Kunstler, those suburban subdivisions created only another geography of nowhere as "the extreme separation of uses and the vast distances between things" hindered the growth of true communities and made people overly reliant on the environmentally damaging automobile. And, as did other New Urbanists, he felt that architectural postmodernism was not helping matters as the starchitects of the era did little more than use irony to cover up the resurrection of an architecture based on "an old-fashioned colonial imperialism." Despite its cleverness and its allusions to history, it remained little more than a decorated version of what had preceded. "Underneath the tacked-on pilasters," Kunstler

concluded, "was the same old box." It was, in other words, modernism in an updated shell.[43]

What could be done to address the loss of community to which such trends had led? Here, Kunstler resurrected the work of Christopher Alexander. Calling for a revival of the pattern language school of thought, he noted approvingly: "[Alexander's] argument implied that the whole built environment consisted not of things at all but only of relationships between and within other relationships." Taken to its logical conclusion, the power of place was thus rooted in the social—rather than the material—realm. Following the lead of Alexander, Kunstler saw successful places as ultimately "based consciously on deep human emotional and psychological needs." Twenty-first-century placemakers would reach this same conclusion.[44]

Kunstler cited the new, master-planned urbanist Seaside development of Andres Duany and Elizabeth Plater-Zyberk, whose uncanny retro-postmodernism was made famous by the film *The Truman Show*, as an example of those doing such important work. Yet, within the realm of New Urbanism, theory, passion, and policy all perhaps best came together in the case of John Norquist. After serving as the Democratic mayor of Milwaukee from 1988 to 2004, Norquist served as CNU president until 2014. In that role, he became a vocal advocate of placemaking, taking part in such things as a discussion on placemaking and revitalization at the 2012 Upper Midwest Planning Conference, held in Madison, Wisconsin, and hosted by the American Planning Association and CNU, and giving talks in places like North Carolina on the topic developing the Cape Fear economy by placemaking.[45]

Norquist described how he put placemaking ideas into action in Milwaukee—and the rationale behind such a move—in his 1998 *The Wealth of Cities: Revitalizing the Centers of American Life*. Strikingly, the book depicted the federal government as an obstacle to urban redevelopment, in the way more often than not. Like Langdon and Kunstler, Norquist saw government intervention in America's cities as a problem that needed to be overcome. Yet Langdon and Kunstler launched their critiques from outside government; here was someone with real political power making this case. And he made it strongly: Norquist began his book at a meeting between members of the US Conference of Mayors and representatives from the federal government following the devastation wrought by the 1992 Los Angeles riots. "In seeking federal aid," he wrote, "we were sending a repulsive message to private capital markets. Imagine the CEO of a private corporation telling the corporation's customers and stockholders that the corporation was out of money and that if it didn't get more money soon it would burn." To Norquist, such as strategy was absurd, but that was what American cities were doing.[46]

Norquist saw the events of the early 1990s as the continuation of a longer history that began with cities going to the federal government for assistance during the Great Depression (the moment of "Original Sin") and continuing through the urban renewal era of the 1940s, 1950s, and 1960s. Here: "The federal government did what it often does when it creates suffering. It grew a new bureaucracy named after the problem it created. In this case it created the Department of Urban Development." In light of such a history, the first step was to "stop viewing cities as problems and governments as the only cure."[47]

This dysfunctional relationship between city and government had created a situation in which city dwellers had become little more than victims. Such individuals no longer felt that they held any power as "remedies intended to revitalize instead overwhelmed the productive assets of cities." Government became the sole driver of redevelopment, and government programs "became bigger and more important than the cities and their people." What was the end result of such developments? "Instead of being rescued, U.S. cities were victimized by federal interference, and their status as victims was institutionalized."[48]

For Norquist, New Urbanism was a way to address this victimization. Like so many before him, he positioned Jane Jacobs as "the first person to resist in any meaningful way the decline of traditional American design." In her campaigns against the state-sanctioned urban renewal efforts that replaced livable neighborhoods with superblocks and high-rise apartment buildings, Jacobs drew attention to the destructive tendencies of government intervention and to the belief that urban spaces needed to foster social interactions of all sorts. It was these ideas, Norquist concluded, that informed the work of the New Urbanists. He came to believe that cities prosper only when, as Philip Langdon put it, "neotraditional development can respond to popular demand by helping restore a sense of community."[49]

One sees the policy implications of such beliefs in Norquist's championing of the rebirth of Milwaukee's downtown RiverWalk. The first RiverWalk segment was built in 1920, next to what was then a Gimbel's Department Store, and was meant to highlight the city's burgeoning commercial district. Yet, during this era, industrial activity still drove Milwaukee's economy, and businesses like tanneries essentially treated the river as their own private sewer. This began to turn around in the 1970s as more stringent water-pollution regulations took effect. In 1997, the removal of a dam directly north of downtown allowed the current to flow faster and generate more oxygen. The Milwaukee River was slowly reborn.[50]

Developers soon took notice. In 1991, the real estate developer Gary Grunau spearheaded the formation of the Milwaukee RiverWalk District (his Schlitz Park business park abutted the river). Two years later, Norquist proposed spending $9.7 million on filling in the gaps in the RiverWalk. He brought on the San Francisco-based architect Ken Kay, a CNU founding member, to design the project. The Milwaukee Common Council ultimately greenlighted the plan, drawing on $7.55 million in city funds and $2.15 million from property owners within the RiverWalk business-improvement district. By the turn of the century, Norquist had expanded the RiverWalk throughout downtown Milwaukee and into the up-and-coming Historic Third Ward neighborhood.[51]

For Norquist, the RiverWalk was meant to revive an underutilized space as a means of decreasing dependence

A wedding inside a repurposed bank along Milwaukee's Riverwalk.

on the automobile and increase walking—and the types of urban sociability thus created. Yet quite soon it became apparent that that attempt to remake a place was also spurring development. The market value of riverfront properties within Milwaukee increased by nearly 150 percent from 2001 to 2013, with an added value of more than $520 million. (During that same period, the total value of all Milwaukee properties increased by 44 percent.) More than half that amount came from condominium projects near the RiverWalk. The creation of this public space also attracted businesses such as Manpower, which relocated from its suburban headquarters to a downtown riverside property in 2007. According to the company spokesperson Mary Ann Lasky, Manpower's new "location on the Milwaukee RiverWalk, with proximity to many restaurants, entertainment venues, hotels and retail establishments, is a very appealing amenity for employees and visitors to our headquarters." By 2015, organizations within Milwaukee calling for increased funding for creative-placemaking endeavors had come to hold up the RiverWalk as emblematic of "the Milwaukee method" for such efforts.[52]

More Eyes on the Street: The Rise of Community Policing

Norquist embraced a Jacobsian approach to planning public spaces for another reason: such a strategy allowed for more eyes on the street and for greater informal surveillance of the urban landscape. He cast himself as a new-style Democratic law-and-order politician. To his mind, the more formal policy of community policing furthered the mission of those like William H. Whyte who sought to rid urban public spaces of undesirables.

Community policing became a nationwide phenomenon after the perceived successes of Rudolph Giuliani, who, after being elected mayor of New York City in 1993, appointed William Bratton police commissioner. Building off the already-ascendant broken-windows theory of policing launched by James Q. Wilson and George Kelling,

Bratton worked to rescue New York's public spaces, spaces that many thought lost to crime and disorder by the early 1990s. The administration connected the presence of smaller-scale nuisance criminals like squeegee cleaners, graffiti vandals, and street beggars to broader fears of urban disorganization, particularly in public spaces shared by city dwellers. Such fears, as the New York Police Department (NYPD) noted in July 1994, could have profound impacts on the city's built environment. They often "cause[d] people to abandon parks" and ultimately "even leave the city altogether." Much of this was due to a perceived growth of violent crime and associated threats to personal safety and property. But there was more to this feeling of decline than a visceral fear for one's life. The NYPD also noted the visual component of urban decline, commenting on "an increase in the signs of disorder in the public spaces of the city," recalling not only Wilson and Kelling but also the work of Kevin Lynch as such incoherent spaces troubled an ordered reading of the urban landscape.[53]

The NYPD also stressed that ordered urban public spaces mattered more than most people thought—and that it was up to the police, with the assistance of conscientious city residents, to provide eyes on the street and surveil such places. First and foremost, safe public spaces fostered the sociability that made urban life so appealing. In a passage that would not have sounded out of place in Jane Jacobs's *Death and Life of Great American Cities* or in a Project for Public Spaces promotional brochure, the NYPD proclaimed: "Public spaces are among New York City's greatest assets. The city's parks, playgrounds, streets, avenues, stoops, and plazas are the forums that make possible the sense of vitality, excitement, and community that are the pulse of urban life." When a city "can't protect its spaces," it needed to address what Giuliani considered the "visible signs of a city out of control." Importantly, Giuliani and others saw such visual disorder as harming the larger health of the city. Surveys conducted in 1992 and 1993 found that 59 percent of people who had moved out of New York recently did so

The New York Police Department's targeted surveillance continues with its SkyWatch towers, which are dropped into hot spots like in the Van Dyke Houses in Brooklyn. The development's residents are nearly entirely people of color.

to improve their quality of life.[54] It was therefore up to the police department to address such an environment, to address issues of visibility by altering what people saw in the city. This could influence more than simply the safety of the city's residents. As the sociologist Alex S. Vitale found: "[Giuliani] argued that by reversing the visible symptoms of social and physical disorder, urban spaces would be economically revitalized." Such policies soon became the norm in a spate of other cities, including San Francisco, Chicago, Los Angeles, Seattle, Baltimore, and Norquist's Milwaukee.[55]

For Giuliani, eyes on the street would create a sense of shared ownership of both the surveillance process itself and the city as a whole. When this happened, as he noted in 1998, "We begin to feel together, we all have a stake in the City. This is what the idea of a civil society is all about." There were, however, many undesirables left out of this vision of the city, and they tended to be people of color. Race quickly became a central component of policing the public spaces of cities such as New York, contributing to the era of mass incarceration that matured during the 1990s. The situation exploded into international headlines in the late 1990s following the tragic police assaults on two Black immigrant New Yorkers, Abner Louima and Amadou Diallo. The resulting political activism focused attention not only on distinctive cases of police assaults and killings but also on the highly racialized "stop, question, and frisk" police practice. The practice was an outgrowth of the department's attention to low-level crime that allowed police officers to detain, question, and partially search a person if there was a reasonable belief that that person was involved or would be imminently involved in a crime. In 2000, the US Commission on Human Rights affirmed what New Yorkers of color had known for years: the practice disproportionately affected them. For example: "88 percent of all stop and frisk subjects in Brooklyn were members of ethnic minorities." People of color constituted 65 percent of the borough's residents. The resulting arrests led to the curious situation in which arrests outpaced the number of reported crimes.[56] The city's approach to creating a more civil public sphere was creating something else altogether. Despite decades of lawsuits, consent agreements, and oversight, these issues infamously remain. For example, the New York Times has found that, from 2008 through 2011, New York police officers issued 2,050 citations for riding a bicycle on the sidewalk in Bedford-Stuyvesant, a predominantly Latinx and African American neighborhood in Brooklyn. During that same three-year period, they issued just 8 citations for the same offense in Park Slope, a primarily white Brooklyn neighborhood. Such policies have dramatically affected the development of the early twenty-first-century city—and produced legacies that mainstream placemakers have failed to address.[57]

From Social Capital to Civil Society

It was no coincidence that Mayor Giuliani referenced the concept of civil society in 1998. Throughout the 1990s, the phrase *civil society* moved from academic into public discourse in a host of cities. But what exactly was civil society, and how best to grow it positively? In 1993, the political scientist Robert Putnam suggested that the answer was to foster social capital. Social capital was not necessarily a new concept, but Putnam brought it squarely into the public imagination. By the early twenty-first century, placemakers—often citing Putnam's work—saw their work as a useful way to foster the social activities necessary to grow social capital. As Putnam explained, social capital was different from but shared a relationship with other forms of capital: "By analogy with notions of physical capital and human capital—tools and training that enhance individual productivity—'social capital' refers to features of social organization, such as networks, norms, and trust, that facilitate coordination and cooperation for mutual benefit. Social capital enhances the benefits of investment in physical and human capital."[58]

Using the case study of regional government in Italy, Putnam concluded that communities that had accumulated social capital found it easier to work together.

Starting in 1970, the Italian state put in place twenty regional government centers across the country. The services offered at each center were virtually identical, but the contexts—economic, cultural, political, and social—in which they took shape differed drastically. Some of these new regional governments proved successful; others failed dramatically.[59]

The question was why. For Putnam: "The best predictor is one that Alexis de Tocqueville might have expected. Strong traditions of civic engagement . . . are the hallmarks of a successful region." Regions that reported less political corruption and more economic growth, including Emilia-Romagna and Tuscany, featured active community organizations that used their regional government centers to engage wide swaths of the population. Over generations, guilds, religious networks, mutual aid societies, cooperatives, choral societies, and neighborhood associations all developed, creating and reinforcing social capital as they went. Regions that failed to address such issues as corruption and economic development adequately lacked such things—and networks tended to be structured hierarchically rather than horizontally.[60]

Putnam's study suggested that social capital had a particular importance in the political arena. Yet the concept touched other aspects of life as well. "The social capital embodied in norms and networks of civic engagement," Putnam continued, "seems to be a precondition for economic development, as well as for effective government." In fact, Putnam saw social capital as driving economic growth throughout much of the developing world—in places such as China and East Asia, where networks based on extended family and close-knit ethnic communities helped "foster trust, lower transaction costs, and speed information and innovation." In the developed world in the 1980s, networking was seen to help with white-collar success. At the same time, industrial districts had the ability to "emphasize networks of collaboration among workers and small entrepreneurs."[61]

Within the United States, Putnam saw the concept of social capital as particularly relevant to troubled cities.

As they did for so many other theorists of the era, the trauma and sheer devastation associated with the 1992 Los Angeles uprisings proved instrumental in shaping his thinking. Writing in the aftermath of this cataclysmic event, he saw in the urban places touched by such violence a distinct lack of social capital. Yes, things like economic disinvestment and white flight mattered. Still: "The erosion of social capital is an essential and underappreciated part of the diagnosis." What this meant was that people of color living in American cities lacked the connections that were valuable in finding employment and that white city dwellers had already learned to cultivate. Thus, the questions that Putnam saw as most important to dealing with such urban crises included: "How is social capital created and destroyed? What strategies for building (or rebuilding) social capital are most promising?"[62]

For Putnam, the beginnings of an answer to such queries involved rethinking social policy. "Classic liberal social policy," he found, "is designed to enhance the opportunities of individuals." However, if social capital was indeed a useful concept, "this emphasis is partially misplaced." In light of such a disconnect, "we must focus on *community* development." More specifically, social capital differed from conventional capital in the fact that it is a "public good . . . not the private property of those who benefit from it." Moreover, it was not something solely provided by private agents. Like such disparate actors as the New Urbanists and Rudolph Giuliani and the NYPD, Putnam saw the need to privilege public good over private concerns at this historic moment. Ultimately, then: "Social capital must often be a by-product of other social activities." Yet again, the impetus for development was placed on the ability to increase sociability.[63]

By the mid-1990s, public intellectuals had taken to this idea of social capital. Fresh from the controversial success of *The End of History and the Last Man* (1992), the political scientist Francis Fukuyama entered into the discussion of social capital with *Trust: The Social Virtues and the Creation of Prosperity* (1995). He immediately made it clear that the book was, at its core, about economics. "But economics," he qualified, "is not what it appears to be either; it is grounded in social life and cannot be understood separately from the larger question of how modern societies organize themselves."[64]

Interestingly, Fukuyama began this discussion on societal organization by noting how societies are *not* organizing themselves. By the mid-1990s, the state was no longer the focal point of organizational efforts. Building off his "end-of-history" thesis, he noted that the triumph of market-driven capitalism had weakened the allure of government interventions of all sorts. "We no longer have realistic hopes," he concluded, "that we can create a 'great society' through large government programs." This lack of faith in the ability of large-scale, state-sanctioned programs to address such ills as poverty and economic development would directly influence the placemaking movement.[65]

In place of the state, Fukuyama saw the development of civil society—which he defined as "a complex welter of intermediate institutions, including businesses, voluntary associations, educational institutions, clubs, unions, media, charities, and churches"—as vital to driving societal organization. Ultimately: "A thriving civil society depends on a people's habits, customs, and ethics—attributes that can be shaped only indirectly through conscious political action and must otherwise by nourished through an increased awareness and respect for culture." Politics took a backseat to other tools used for things like economic development.[66]

But what did Fukuyama mean by *culture* and its relationship to economy? Here, culture seemed to be the means of growing sociability and the human relationships at the heart of sociability. To Fukuyama, "the economy constitutes one of the most fundamental and dynamic arenas of human sociability" as there are few types of economic activity that do "not require the social collaboration of human beings." And, while people worked in organizations to satisfy their individual needs, the workplace also drew them out of their private lives and connected them to a wider social world. Work was, therefore, more than economic; it was social.[67]

Yet: "American society is [in the 1990s] becoming as individualistic as Americans have always believed it was." This led directly to declines in trust and community, the main ingredients of human sociability. What was needed were ways to reinvigorate a weakened civil society. This meant amplifying the concept Putnam trumpeted: social capital. To Fukuyama, social capital, "the crucible of trust and critical to the health of an economy, rests on cultural roots." This meant that an emphasis on culture was, indeed, warranted as sociability and social capital were crucial first ingredients of economic growth.[68]

Fukuyama was writing at a time of pronounced economic stability in the United States: the Cold War was over, and the nation was experiencing tremendous financial growth. In such an atmosphere, social capital could be called on to do more than facilitate such economic growth. It could now be used to build a better-balanced civil society, one that finally got right the relationship between the individual and the communal. Simply put, social capital moderated the destructive tendencies of American individualism. "The United States," Fukuyama noted, "is heir to two distinct traditions, the first highly individualistic, and the second much more group and community oriented." By the late twentieth century, the "challenge for the United States" became "bringing these tendencies into better balance." Whereas those within the New Urbanist movement saw design as central to this mission, Fukuyama saw social capital as vital. And, with the defeat of communism and the subsequent end of history, the timing was right. As Fukuyama ended his book: "Now that the question of ideology and institutions has been settled, the preservation and accumulation of social capital will take center stage." Placemakers would, in the early twenty-first century, work to build that stage.[69]

All this came together in Robert Putnam's magnum opus, *Bowling Alone: The Collapse and Revival of Ameri-*

can Community (2000), which elaborated on his earlier work. Putnam continued to stress the importance of social capital—and to flesh out its definition. Perhaps in light of Fukuyama's contribution, but also with his earlier work in mind, Putnam now saw social capital as "connections among individuals—social networks and the norms of reciprocity and trustworthiness that arise from them."[70]

Much of Putnam's attention throughout *Bowling Alone* was on American cities. Perhaps not surprisingly, Putnam traced his understanding of social capital back to Jane Jacobs as he saw this as a concept "of which she is one of the inventors." More specifically, he honed in on her *Death and Life of Great American Cities*: its understanding of social contact and its relationship to the concept of social capital. According to Putnam: "[Jacobs] argued that where cities are configured to maximize informal contact among neighbors, the streets are safer, children are better taken care of, and people are happier with their surroundings." This type of social contact ultimately "developed a sense of continuity and responsibility in local residents."[71]

Putnam quoted at length from Jacobs's formative work as he stressed the casual, informal social contact of the city, "most of it fortuitous." Echoing Jacobs, he found that such contact was "metered by the person concerned and not thrust upon him by anyone." It allowed for the development of "a web of public respect and trust to develop," one that came to serve as a valuable resource for city dwellers, particularly during times of need. Yet this contact also allowed for informal surveillance to take place. "Higher levels of social capital," he concluded, "translate into lower levels of crime."[72]

What, then, was social capital, ultimately? "Merely new language for a very old debate in American intellectual circles." Here, Putnam again saw the influence of Jacobs. Jacobs's conception of social contact was rather superficial, as valuable interactions that often took place between urban residents who did not even know each other's names came to influence those who came after

her. As we will see in chapter 3, contemporary place-making advocates such as Richard Florida continue to sing the praises of such informal social bonds. Yet Putnam himself seemed to be pushing for something deeper, as is evident in his emphasis on the distinction between bridging social capital, which connected people across a social divide, and bonding social capital, which strengthened existing relationships. Bridging social capital was what should be worked toward as it "can generate broader identities and reciprocity." Showing a clear debt to Fukuyama, Putnam found that "community has warred incessantly with individualism for preeminence in our political hagiology." Once again, social capital—and particularly bridging social capital—was seen as the way to strike a useful balance between individualism and community as it essentially prompted people to connect with those unlike them. After all, the very idea underlying bridging social capital was that subcultural and other differences exist but that those differences could be meaningfully connected to produce mutually beneficial action. But is bridging or bonding the way to go? These often-ambiguous relationships—and the questions they provoke—continue to challenge placemakers in the twenty-first century.[73]

Yet Putnam was quite clear on who benefited most from access to social capital: underprivileged communities in American cities. Echoing his earlier work, he once again stressed that the decline in activities associated with social capital in urban communities, including "community monitoring, socializing, mentoring, and organizing," was a crucial component of "the inner-city crisis, in addition to purely economic factors." Again echoing Fukuyama, he argued that the accumulation of social capital could create wealth and lead to economic rebirth. Social capital did more than create livable urban spaces; it also helped the inhabitants of those spaces succeed economically. "A growing body of research [and here he cited Fukuyama's *Trust*]," Putnam reported, "suggests that where trust and social networks flourish, individuals, firms, neighborhoods, and even nations prosper."[74]

Bowling Alone concluded with Putnam formulating plans for an environment that would foster the development of social capital. What he called for sounded a whole lot like a nascent form of placemaking. While he saw some value in New Urbanism and its use of built form to facilitate the possibility of sociability and interconnection, he called for a type of urbanism that better recognized the power of culture. The New Urbanists' focus on design undoubtedly helped create spaces where social capital could take root successfully. But such designers did not focus on *how* social capital would then flourish in such well-designed spaces. To Putnam, "the arts and cultural activities" were the keys to bridging this relationship between place and social capital: "Art manifestly matters for its own sake, far beyond the favorable effect it can have on rebuilding American communities. Aesthetic objectives, not merely social ones, are obviously important. That said, art is especially useful in transcending conventional social barriers. Moreover, social capital is often a valuable by-product of cultural activities whose main purpose is purely artistic."[75]

Usefully, Putnam offered two places where he believed arts and cultural activities fostered social capital: Liz Lerman's Dance Exchange Shipyard Project, which attempted to heal the wounds created by the closing of the Portsmouth, New Hampshire, shipyard, and Chicago' Gallery 37, which "provide[d] apprenticeships for young budding artists" so that they could "follow their own muses, building social connections among artist-mentors, artist-apprentices, and observers." Within such places, the arts were actively creating social capital that was working "to restore American community for the twenty-first century through both collective and individual initiative." Such would be the logic that came to inform placemaking as the new millennium began.[76]

Yet the examples provided by Putnam were instructive in another way: they were located in almost exclusively white spaces. Portsmouth's population was over 90 percent white during the 1990s, and Gallery 37 was rooted in the predominantly white Chicago Loop—in a city where over 60 percent of African Americans lived in communities that were 95 percent Black. While the port performance was a temporary project that ended over twenty years ago, Gallery 37 grew beyond this early limitation. Particularly since the turn of the century, the program has dramatically expanded; in 2000, the array of programs offered at Gallery 37 was centralized as After School Matters. What began as a way of bringing around 250 students into the heart of the city has developed into a robust set of after-school programs serving thousands of students from all backgrounds in summer and enrichment programs throughout the city, including the Loop location of Gallery 37. Yet the bulk of the work of After School Matters came to be done in sites located throughout high-poverty communities, particularly on Chicago's West and South Sides. Here, participating students outperform their peers on many development measures, including college attendance, and the very assessments one would expect to find in vibrant networks of social capital: feeling supported, engaged, and safe. Two decades of funding and attention provided the capacity to use placemaking to develop social capital and thus create new opportunities for some of Chicago's most vulnerable residents.[77]

At the turn of the century, however, many programs were not as successful in creating the social capital that Putnam had seen as so important for urban redevelopment. Moreover, outside the placemaking discourse, there was a growing belief that this limitation had real implications for economic outcomes, particularly for American cities still suffering the ravages of deindustrialization. As early as 1987, in his *The Truly Disadvantaged: The Inner City, the Underclass, and Public Policy* the acclaimed sociologist William Julius Wilson highlighted the relationship between a lack of broad social networks and economic well-being. As he wrote on "the problem of social networks," those living in "highly concentrated poverty neighborhoods in the inner city today not only infrequently

Wilson's landscape of geographic and social isolation in Chicago.

interact with those individuals or families who have had a stable work history"; they "also seldom have sustained contact with friends or relatives in the more stable areas of the city or in the suburbs." Residents of economically and geographically segregated African American communities may have been developing social capital, but it was only among people in similar structural positions. This social isolation had dire repercussions, a point that Wilson elaborated on in 1996's *When Work Disappears: The World of the New Urban Poor*. Wilson still saw a lack of "social positions" and "networks of social relationships" as integral to inner-city poverty. By the mid-1990s, he emphasized that such networks had been further weakened by "fundamental structural changes in the new global economy." How, he asked, were residents of the South Side of Chicago supposed to adapt to such a profound worldwide economic restructuring, particularly when they were increasingly marginalized? While Wilson's work would influence the replacement of public housing projects that concentrated poor residents with new mixed-income developments, as the economy expanded the same ideas would not extend into the placemaking discussion. The question would not come to the fore for more than a decade, only after the Great Recession.[78]

Across the country, antiglobalization activists built a movement meant, in part, to speak to such developments, which flew under the radar of most mainstream observers. Such efforts exploded into the international spotlight in 1999 at the World Trade Organization (WTO) ministerial conference. The radical nature of this movement—news coverage focused on anarchists and masked individuals smashing the windows of Starbucks franchises—meant that it really did not affect mainstream understanding of placemaking. Yet many of these activists stressed that they were part of a broader anticapitalist movement, one that stretched at least from the 1969 Weather Underground "Days of Rage" attacks on multinational corporations to anti–North American Free Trade Agreement protests at the 1996 Democratic national convention in Chicago.

Such voices came to question what property should be used for, with one group of anti-WTO protestors—the ACME Collective—articulating that "personal property" is "based upon use" while "private property is based upon trade." In such a system: "The premise of personal property is that each of us has what s/he needs. The premise of private property is that each of us has something that someone else needs or wants." Thus, free trade advocates sought to ignore the common good of public spaces and put in place "a network of a few industry monopolists with ultimate control over the lives of everyone else."[79]

It would be folly to suggest that this critique was broadly influential at the time, though challenges to the sanctity of private property would become more prominent after the housing bubble collapsed eight years later. Yet protests like the ones that unfolded in Seattle in 1999 did elevate issues of economic inequality, so much so that even those like Renato Ruggiero, the director-general of the WTO from 1995 to 1999, reached the conclusion that "the WTO cannot operate in isolation from the concerns of the world in which it exists." Instead, organizations and individuals alike had to confront the "democratic deficit" that was emerging in the global economy and in economies across the world, a move that would address the "concern that the trading system is out of touch with the very people it was designed to serve" and whose lives it was meant to improve. For Ruggiero as well as others, responding to the democratic deficit meant local action as much as global treaties; in the twenty-first century, it would be up to people to define the economies in which they would live and work.[80]

Organizations like After School Matters can thus be seen as crafting a template for how creative placemaking would grow in the new century. The site-specific, community-oriented approach to programming of After School Matters foreshadowed the neighborhood-centric model of placemaking that would inform the practice in the early years of the twenty-first century. Moreover, its emphasis on fostering public/private partnerships—its partners include the Chicago Park District, the Chicago Department of Family and Support Services, the Robert Wood Johnson Foundation, and companies like Nationwide Acceptance, LLC—also came to be a defining characteristic of placemaking in the twenty-first century. Yet, perhaps most importantly, the group came to see its work as creating real economic value for young people of color throughout Chicago. Through an innovative apprenticeship program, students could receive paid positions not only in arts-related fields but also in other creative fields like video editing, music production, and architecture.[81] By the early twenty-first century, myriad American cities would place such relationships and programs at the epicenter of their creative-placemaking endeavors.

CHAPTER 3

Into the Twenty-First Century

(*Previous spread*) Street food at a night event in Detroit's Eastern
Market, an acclaimed creative-placemaking project.

A Return to the Third Place: Placemaking in the Early Twenty-First Century

By the early twenty-first century, many observers came to note the disintegration of the social ties that went along with Robert Putnam's bowling-alone thesis. It was with this atmosphere in mind that Ray Oldenburg put out *Celebrating the Third Place: Inspiring Stories about the "Great Good Places" at the Heart of Our Communities* (2001). This work, a collection of 19 essays written by proprietors of successful "Great Good Places," revisited the concept of the third place that Oldenburg had originally introduced in his 1989 work, which had been reissued in 1999 by Marlowe and Company. For an unnamed reviewer writing in *Publishers Weekly*, the book could have very well been subtitled "Bowling in Groups." To *Celebrating the Third Place* editorial director Matthew Lore, the book was in fact an antidote to Putnam's recently published *Bowling Alone*, one that "celebrates places where people can come together for good of community and democracy."[1]

Oldenburg's 2001 case studies illustrated how placemaking was coming to be viewed in the first decade of the twenty-first century. The places chronicled in this turn-of-the-century work highlighted how such sites became successful third places. Once again channeling Tocqueville, Oldenburg saw the third place as, first and foremost, leading to greater social association. Yet there was nothing official about such sites; the key to them was their informality. It was therefore up to the participants to navigate them, without any sort of preprogrammed guidance or assistance. To Oldenburg, these spaces suggested "the capacity of Americans to do what needs doing without depending on government." Politics and public policy, in other words, were not the answers.[2]

At their most basic, then, these were places "in which people relax in good company and do so on a regular basis." Of course, there must be a physicality to such places, and Oldenburg was heartened by the positions on architecture and planning staked out by the New Urbanists during the 1990s. "But," he ultimately asked, "is the archi-

tectural remedy sufficient?" The answer he offered was a resounding no. What happened in such places mattered just as much as—if not more than—how these sites were designed. Practicality, in other words, trumped aesthetics.[3]

Yet there was a sense that these third places did something more than provide a place where Americans could kick back with some friends. If such sites could promote sociability, they also provided room where one could exert a sense of individual agency. One felt in charge of one's destiny in a good third place. "Our society, alas," continued Oldenburg, "has become much like Tocqueville's homeland, in which governmental agencies are expected to do whatever needs doing." Yet this was the wrong approach as "what government does is done remotely and impersonally." Such an impersonal and distant relationship brought about a situation in which "its focus on our weaknesses and dependencies and its policies define us accordingly." The third place could empower individuals to move closer to some sort of independence.[4]

Interestingly, the third places that Oldenburg highlighted were almost all part of the food and service industries, with the majority being small coffee shops and taverns rooted in specific neighborhoods: Civilization Coffeehouse (Cleveland), the Neutral Ground Coffee House (New Orleans), Tunnicliff's Tavern (Washington, DC), and the Blue Moon Tavern (Seattle). Third places apparently do not lack for caffeine or alcohol. In fact, almost every site highlighted by Oldenburg was a site of economic exchange. For Civilization founder Bob Holcepl, the economic component flowed from the other aspects of the endeavor; there was a reciprocal relationship between his individual financial improvement and the communal aspects of Civilization. On discovering Oldenburg's *The Great Good Place*, Holcepl noted: "I realized there was a name for the type of establishment we were building.... Now we had an even greater mission in addition to restoring a building, creating a business, and helping revitalize a neighborhood—we were taking part in the resurrection of a form of social interaction that was almost extinct in America."[5]

Patrons and staff alike relax in Johnsen's Blue Top Drive-In, a prime example of a community-centered restaurant.

And this increased social interaction brought more customers, which helped grow Holcepl's business. Increased business helped improve the space itself, which meant more social interaction and choices for customers. With these additional profits, the coffeehouse was able to improve lighting within the space, install air-conditioning, and expand its menu to include lunch items "to better serve [its] expanding customer base." After a few years, Holcepl was able to purchase the building and renovate its exterior, a renovation job that included a historically correct reconstruction of parts of building that had been destroyed over time. Such developments contributed to the well-being and attractiveness of the Cleveland neighborhood that housed Civilization while allowing Holcepl to remain firmly in control of his position as an independent small-business owner. He was even able to hire a few more employees.[6]

Throughout the early twenty-first century, such economic potential came to be stressed in the evolving discourse on creative placemaking. As a significant part of this process, the word *development* has, perhaps not surprisingly, increasingly attached itself to placemaking efforts. On the one hand, this focus on development can be seen as a logical end point to efforts to reconcile the relationship between individual and community. A successful placemaking endeavor can develop both sides of this relationship and, in the process, work to reconcile the tensions between them. Here, the language of community development is employed to discuss the way urban places become more vibrant and livable.

Yet, as American cities continued to hone their postindustrial identities in the early twenty-first century, the practice of placemaking allowed for a different understanding of community development to become attached to this concept. Here, the emphasis has been on the types of financial activity that placemaking can bring to a community. This has particularly been the case following the Great Recession as cities have had to address unemployment, mass foreclosures, capital mobility, and loss of revenue for even basic city services. Placemaking thus provides a new strategy to deal with a changing economic landscape. The creation of creative places attracts both people and investment, and this, in turn, leads to job creation. To advocates of this process, placemaking projects offer a way to reconcile the tensions between community health and individual well-being further, though now this relationship is viewed primarily through the lens of economics. The city remains the place where this can happen, with a renewed emphasis on the neighborhood as the core unit of space in placemaking. Above everything else, the neighborhood becomes a physical space for the economic mobility of individual residents.

Such ideas have proved particularly attractive to governments at all levels struggling to maintain funding for even basic services. They have proved similarly appealing to foundations, companies, and arts-based organizations who also struggled to find their footing in a postrecession economy, though many of the issues that these actors face often have even deeper historical roots. As growing numbers of such actors embrace placemaking, the concept itself has become more mainstream. On one level, this has allowed placemaking to become a multi-million-dollar industry. Yet the mainstreaming of placemaking has privileged the financial activity that such projects can bring while downplaying the ways such work can speak to issues of politics, race, and social justice.

This chapter will focus on scholars whose ideas have provided the intellectual scaffolding for the explosion of creative placemaking in the early twenty-first century. Yet someone needed to put these ideas to use: nonprofit organizations like Project for Public Spaces (PPS) and government agencies such as the National Endowment for the Arts (NEA), often working directly with a litany of academics and planning experts, have provided the organizational infrastructure necessary for the actual completion of placemaking projects. At the same time, myriad corporations, philanthropic organizations, universities, government leaders, and city planners have offered additional support for many placemaking endeavors. Strategic partnerships between such disparate actors have been formed, allowing for larger projects, reliant on various funding streams, to be undertaken. This process culminated in the creation of ArtPlace America, a collaboration among a number of foundations, financial institutions, and federal agencies that uses creative placemaking as the primary tool for comprehensive community planning and development. Since its inception in 2011, ArtPlace America has invested over $50 million in placemaking projects across the United States. By the end of the 2010s, there was little doubt that creative placemaking had become a growth industry.

The Creative Class and the City: Richard Florida and the Mainstreaming of Placemaking

Richard Florida is one of the most important influences on contemporary-placemaking endeavors. Even as many have come to question his work, his shadow continues to loom large. Florida—who has a doctorate in urban planning from Columbia University and has taught at such institutions as the University of Toronto, New York University, and Carnegie Mellon University over his thirty-year career—has explicitly worked to synthesize the writings of those who came before him; he does not hide the fact that he sees himself as building on books written by the likes of Jane Jacobs, William H. Whyte, and Ray Oldenburg and groups such as PPS. Like them, Florida has concerned himself with what makes cities thrive. And, like Jacobs and her acolytes, he sees the arts as important to this process. Yet what is perhaps most noteworthy is that, very quickly and more than others, he connects a sense of well-being to primarily economic concerns. "The reason I came to arts, culture, and diversity issues," he has honestly explained, "is simply because I found them to be fundamental to the process of economic development."[7]

While Florida has written a host of books that have

inspired would-be placemakers, his most influential remains *The Rise of the Creative Class... and How It's Transforming Work, Leisure, Community, and Everyday Life* (2002). For Florida, as for others before him such as Daniel Bell and Peter Drucker, there has been a global shift from an economy based on manufacturing toward one based on information and knowledge. Yet taking such an analysis a step further, Florida has asserted: "We now have an economy powered by human creativity." In fact, such creativity has emerged as "the most highly prized commodity in our economy." Within this "creative economy," things like knowledge and information "are the tools and materials of creativity."[8]

By the early twenty-first century, workers able to harness the potential of human creativity were becoming more and more common. Harking back to the work of William H. Whyte, Florida found: "Like Whyte's managerial class, which 'set the American temper' in the 1950s, the Creative Class is the norm-setting class of our time." Yet, where the organization man was marked by homogeneity and conformity, the creative class embraced "individuality, self-expression and openness to difference," among other things. For Florida, the creative class traced its roots back to Jane Jacobs, who, he says, once argued that "the creative environment . . . required diversity, the appropriate physical environment and a certain kind of person to generate ideas, spur innovation and harness human creativity." Such people could come together in "a setting where difference, nonconformity and creativity could thrive."[9]

As such comments suggest, place mattered for Florida, particularly as "it is geographic place rather than the corporation that provides the organizational matrix for matching people and jobs." And, in this new world, it is no longer such organizations that define us: "Instead, we do this ourselves, defining our identities along the varied dimensions of our creativity." Again, a balance between community and individuality must be struck. Or as Florida posed it: "A central task ahead is developing new forms of social cohesion appropriate to the Creative Age." One can see why such ideas would appeal to later placemakers; this is what placemaking strives to do.[10]

And what types of places do these people like? They were drawn to places where "many outdoor activities are prevalent." More specifically, the creative class appeared to be drawn to the urban street scene, as it "is social and interactive." This is not to suggest that people have not always looked for social interaction within their communities. "But a community's ability to facilitate this interaction," according to Florida, "appears to be more important in a highly mobile, quasi-anonymous society." For him, then, these spaces needed to be participatory. Citing Ray Oldenburg's work, Florida found that the creative class was concerned with "*quality of place*" and that its members "want to have a hand in actively shaping the quality of place in their communities." In other words, they *want* to create their own Great Good Places. Creative placemaking would provide the vocabulary and the means to make this happen.[11]

Importantly, the creative class had the resources to help such processes along. Such individuals no longer had to worry about money. Speaking on the freedoms afforded to those living in this postscarcity economy, Florida cited the work of Ronald Inglehart, who has found a "shift from emphasis on economic and physical security toward increasing emphasis on self-expression, subjective well-being, and quality of life." Such a shift would not matter if there were not so many Americans taking part in it. It was true that, in the early twenty-first century, the creative class was still a statistical minority within the American labor force; Florida placed 38.3 million American workers, or close to 30 percent of the US workforce, within this group. At the same time, the working class constituted 33 million workers (a quarter of the workforce), while the service class included 55.2 million workers (43 percent of the workforce). Yet, writing in 2002, Florida saw the creative class clearly growing in both numbers and importance. According to him, the working class "no longer has the hand it once did in setting the tone or establish-

ing the values of American life," while the service class lacks the economic clout to replace them. The creative class is therefore the future; their interests will drive the shaping of the American economy and American culture from now on.[12]

As 2002's *The Rise of the Creative Class* suggested, this group was drawn to cities, a fact Florida explored in 2004's *Cities and the Creative Class*. Here, he really delved into the relationship between creativity and economic growth and how this relationship drove economic activity in urban centers, which in turn drove national economic growth. Once again, he based his thoughts on the work of Jane Jacobs, approvingly noting: "Jacobs called attention to the role of creativity and diversity as engines for city growth. She saw the significance of eclecticism and inventiveness as important components of city life." In fact, he himself found that "the role of culture is much more expansive" than even Jacobs had suggested and that "the key to economic growth is to enable and unleash that potential."[13] This argument was further developed in 2008's *Who's Your City? How the Creative Economy Is Making Where to Live the Most Important Decision of Your Life*, in which, much like Jacobs and Whyte before him, Florida argued that "economic activity—such as trade, commerce, and innovation—has always originated in cities." In many ways, *Who's Your City?* simply echoed the arguments offered within *Cities and the Creative Class*. Yet *Who's Your City?* furthered developed the claim that something needed to spur the movements of human capital that made such urban economic growth possible—what Florida called *clustering*. And the process of clustering inevitably led to even more economic growth. Within cities that encouraged and fostered such movement, he noted, "clustering makes each of us more productive, which in turn makes the place we inhabit even more so—and our collective creativity and economic wealth grow accordingly." The city thus became the engine of economic growth for the entire nation. This concept is one that placemakers in the twenty-first century have truly taken to heart.[14]

But what do such creative cities actually look like? And how do they come to encourage clustering? Florida began to answer such questions in *Who's Your City?* He found that place could have a strong impact on both individual and communal activation. Quality places "encourage people to do more than they otherwise would, such as engage in more creative activities, invent new things, or start new companies—all things that are both personally fulfilling and economically productive." Referencing the work of his collaborator Irene Tinagli, Florida wrote that this type of activation came from "the visual and cultural stimulation that places can provide"—more specifically, places like parks and open spaces and cultural offerings. This created a "regenerative cycle" in which such stimulation "unleashes creative energy, which in turn attracts more high-energy people from other places, resulting in higher rates of innovation, greater economic prosperity, higher living standards, and more stimulation." As the twenty-first century continued, such ideas came to inform the formula for many placemaking efforts.[15]

This focus on the environment's impact on individual behavior suggested that, despite the concerns voiced by communitarians and others, individuality remained important in the creative age. In fact, the people whom Florida interviewed actually welcomed the way the city allowed you to be yourself—and sometimes by yourself—and "rarely wished for the kinds of community connectedness [Robert] Putnam talks about." In fact, "creative types" and "bohemians" had historically tried to flee such confining places. This did not mean that such individuals did not see value in community. However: "They did not want it to be invasive, or to prevent them from pursuing their own lives. Rather, they desired what I have termed 'quasi-anonymity.'" Here, the environment must serve to inspire the individual, not necessarily help a community cohere.[16]

Such an emphasis led Florida to fixate on bohemians, or the artists who make a city creative. Why did their existence matter? "The findings here," he announced, "sug-gest a close association between bohemian clusters and high-technology industry." And which cities did well on his "Bohemian Index," which was "essentially a ratio that compares the percentage of bohemians in a region to the national pattern" (seriously)? Portland, Oregon, New York, Seattle, Los Angeles, San Francisco, Boston, and Austin, Texas, all scored well. And which cities fared poorly? Cleveland, Albany, Pittsburgh, Buffalo, and Baltimore. *What* these bohemians produced did not really matter. Instead, what struck Florida as important was that their presence seemed to inspire the creativity of those driving tech firms. To oversimplify such a relationship, art equaled urban economic growth.[17]

You might think that the Great Recession would lead to Florida rethinking some of these claims, but you would be wrong. In 2010, he published *The Great Reset: How New Ways of Living and Working Drive Post-Crash Posterity*. Despite the severity of the 2008 economic crisis, he saw little good in "government bailouts, stimuli, and other patchwork measures designed to resuscitate the old system." "Government spending," he concluded, "simply can't be the solution in the long run." Why not? The concept that the public sector could serve as a spur to economic development was an idea rooted in the twentieth century. In the early twenty-first century, government at all levels "simply lacks the resources to generate the enormous level of demand needed to power sustained growth."[18]

In fact, Florida saw the Great Recession as proving that he was right all along—it is just that no one kept up with his thinking. All those lost jobs and foreclosures were just evidence that the old system was dying right in front of our eyes. With such an idea in mind, Florida was able to recast the housing crisis as an important corrective to the standard belief that homeownership was a marker of economic success and stability. In a new creative economy that "revolves around mobility and flexibility," owning a home can, in fact, become an albatross, "an economic trap, preventing people from moving freely to economic opportunity." As the American economy continued to evolve, homeownership would not be necessary: "We'll be spending relatively less on the things that defined the old way of life."[19]

Despite this move away from homeownership, there remained something spatial to this "reset economy." According to Florida: "Every major economic era gives rise to new, distinctive geography of its own." The collapse of 2008 marked the end of the suburban model of economic growth while highlighting the continuing decay of certain types of urbanism. As for the latter, the crisis showcased a Detroit that was crippled "by its very model of sprawling, auto-dependent growth." The new economy would run "more on brains than brawn," with the result that "there [would] surely be far fewer manufacturing jobs in America after the crisis." Creativity, rather than size or strength, was now driving economic growth.[20]

Yet the suburban model was just as obsolete, and the collapse illustrated that such a model of development could not "channel the full innovation and productive capabilities of the creative economy." The suburbs—those places that represented wealth for so much of the twentieth century—were boring; the places that would drive postcrash economies were those with "the highest velocity of ideas, the highest density of talented and creative people, and the highest rate of metabolism." "Velocity" and "density" were not words that many people used when describing the suburbs. For Florida, the recession drove this point home with a vengeance.[21]

Ultimately, then, the Great Recession was just proof that Florida's approach to urbanism was right all along. The city was once again the prime engine of national economic development, and everyone just had to give the creative types space and let such places evolve organically. Municipal governments had to loosen up to assist with this process; Florida called on city leaders to embrace "liberal zoning and building codes within cities, to allow more residential development and more mixed-used development." Yet that was to be the extent of public-sector

involvement in any sort of economic recovery strategy; it was leading by getting out of the way. Or as Florida forcefully concluded: "One thing is for certain: government is not the prime mover in Great Resets."[22]

Yet arts and culture *were* prime movers of financial growth, a fact that Florida believed would become more apparent following the economic crash of 2008. "Arts," according to Florida, "[had become] an important component of the creative economy engine." There was a symbiotic relationship here as the economy "benefits from considerable spillovers and synergy as art and design expertise combines with technological know-how, producing all kinds of innovative new goods and services." Increasingly, cities would serve as the spaces where this relationship would flourish, a development that would attract creative-class members to them. By 2010, Florida could conclude only that the importance of art and design was leading to a situation where "new patterns of living and working gradually take shape and begin to remake the economic landscape."[23]

Such ideas have come to influence a host of placemaking efforts. In addition to his countless speeches and many books, in 2005 Florida founded the Creative Class Group, a global advisory services firm meant to help clients harness the emerging power of the creative class. Clients have included such corporations as Google, Microsoft, and BMW; cities such as Seattle, Austin, and Tampa Bay; philanthropic organizations like the John S. and James L. Knight Foundation; and such government bodies as the US Department of Labor and the United Nations.

What Is Old Is New Again: The Continued Relevance of PPS

If Richard Florida provided the intellectual kick start for placemaking in the early twenty-first century, then PPS provided the organizational apparatus to help the concept reach new audiences and new levels of influence. Because of its long history, PPS had the infrastructure in place to take advantage of the moment. There is little doubt that Florida and PPS came to have a mutually beneficial reciprocal relationship in the early twenty-first century. Each shared a profound debt to the earlier work of such urban theorists as William H. Whyte and Jane Jacobs. Moreover, PPS publications like Cynthia Nikitin's 2013 "All Placemaking Is Creative: How a Shared Focus on Space Builds Vibrant Communities," for example, used Florida's academic treatises to support their real-world projects, while Florida recognized in PPS a group that had been doing the work he saw as important for the growth of cities for close to forty years.[24]

In 2000, PPS published *How to Turn a Place Around: A Handbook for Creating Successful Public Spaces*, in which it noted: "This book is not about any particular kind of community. It is about every community, because we believe that the same principles apply [to myriad locations]." It argued that "each community has the means and the potential to create its own public places," particularly when drawing from a PPS-supplied tool kit.[25]

In line with Richard Florida, the book saw public places helping give areas identities while benefiting them economically; in New York City, land values near Bryant Park soared after its activation, while the Greenmarket on the north end of Union Square served as a catalyst for greater investment in the surrounding community. "What," the authors asked, "makes a place great?" On the basis of research conducted at over one thousand public spaces around the world, the authors identified four qualities of successful public spaces: "Accessibility, Activities, Comfort, and Sociability."[26]

Yet what seemed to preoccupy the authors more was what made a public space fail. On the one hand, such spaces were less than great when they were "empty, vandalized, or used chiefly by undesirables." This often happened when those behind such spaces employed a "project-oriented approach to planning," as most American cities do. A project usually came about because of a political situation, a budget, or the work of an urban planner or developer. This discipline-based approach relied too heavily on experts who had never had to live with the spaces they created: "The professionals responsible for activities that directly affect public spaces—such as planning, traffic and transit, recreation, and education—have roles that are peripheral to those spaces." Moreover, this expert-driven planning usually just brought about the need for more money, which usually is not there. In a chapter titled "Money Is Not the Issue," the authors wrote: "The money is usually just handed to a professional designer, who, it is supposed, 'knows better.'" In fact, when money drives the development process, "this is generally an indication that the wrong concept is at work—not because the plans are too expensive, but because the public doesn't feel like the place belongs to them."[27]

At the same time, government seemingly had little role to play in the creation and maintenance of great public spaces. Simply put, the authors found: "Government is compartmentalized. As a result, people working in government are limited in their ability to deal with public spaces effectively." The end results of such a compartmentalization process were the creation of "bureaucratic roadblocks" and "narrowly defined responsibilities." No one in government saw himself or herself as the steward of such sites, so they often went unnoticed or underutilized by the public sector. Moreover, like their counterparts in the professions, government officials rarely had to live with the spaces that fell under their jurisdiction. It was little wonder that they had little interest in getting them right.[28]

Instead, great public spaces came only "out of a discussion with a community about how to create good public places within that community." Rather than being project or expert driven, such spaces needed to be community based and place driven. There can be no preplanned design for them; instead, all parties involved must cooperatively "develop a vision" for the project. Finally, such projects should take advantage of the benefits of triangulation, which "means locating elements in a way that greatly increases the chances of activity occurring around them": "The idea is to situate them so that the use of each builds off the other." Richard Florida would also pick up on this concept a few years later. Yet, more importantly, this was exactly what William H. Whyte was calling for, and in exactly the same language, in 1980's *The Social Life of Small Urban Spaces*.[29]

This does not mean that PPS was simply rehashing the past at the turn of the century. As noted in chapter 1, initially it had difficulty convincing municipalities and funders to buy into its approach to urban redevelopment. Buttressed by the support provided by the works of Florida and others (who themselves had been influenced by its pioneering work and that of Whyte), it was finally able to convince clients that it had been right all along. By positioning itself as the liaison between space, individual, and community—as the mediator in the discussion that must be behind the design and implementation of such spaces—it was asserting its importance and role as an organization. And, by calling for something beyond government—indeed, beyond urban planning—it sought to carry out such a role utilizing a new vocabulary. By the early twenty-first century, it was employing the language of placemaking. "Placemaking," the authors of *How to Turn a Place Around* concluded, "is about doing more than planning." Whereas many large-scale urban redevelopment plans were simply "too big, too expensive, and simply take too long to happen," placemaking provided opportunities for shorter-term, smaller-scale projects to take root. These projects could be a way of "testing ideas" while at the same time "giving people the confidence that change is occurring." Moreover, the public nature of these types of projects, along with the fact that testing involved community voices in the planning process, drove home the point that, when it comes to residents' involvement in such projects, "their ideas matter."[30]

Such ideas were further developed in "Public Markets as a Vehicle for Social Integration and Upward Mobility," a 2003 report commissioned by the Ford Foundation and

authored by PPS with the assistance of the nonprofit organization Partners for Livable Communities. The report immediately took issue with the "big-project" approach to urban renewal and, more specifically, called out the historical overreliance on highway, office complex, and convention center construction. When it came to economic payoff, such projects "have tended to produce more menial jobs than meaningful ones." Most distressingly: "They have done serious damage to the urban fabric—its streets, parks, and public spaces—which, in turn, leads to further and even downward spiral."[31]

Yet, in the early twenty-first century, practitioners of urban revitalization across all sectors were beginning to see the need to rethink large-scale solutions to urban problems. Groups like PPS had come to "understand the importance of public spaces that connect everything together" as a means of healing the ways previous efforts tended to isolate city dwellers. There was also an economic component to such an approach as it had become clear that "public gathering spaces are inextricably related to the potential for economic opportunity and upward mobility of lower-income people."[32]

For PPS, the ideal form of such a public space was the public market. Public markets were one "of the most obvious, but perhaps least understood, methods of enhancing social integration in public spaces and encouraging upward mobility." On the one hand, they promoted sociability; they brought communities together in a tangible, well-defined place. But they also addressed the needs of the individual. As the report continued, public markets also "create a sustainable vehicle for upward mobility and individual empowerment for low-income communities." And while "public markets must have public goals," the economic component of such spaces could not be overlooked.[33]

But the key to such economic potential, the report concluded, was its inclusive public component. By examining the workings of eight markets across the United States—including the Berkeley Flea Market, the Findlay

Market (Cincinnati), and the RFK Stadium Farmers' Market (Washington, DC)—the report found that these spaces work best when they are seen as "social places." Here, design mattered: "The data appear to show that social integration is achieved by a market that has 'something extra.' The market must have an attractor beyond its role as a place to buy goods." This was what placemaking was meant to do.[34]

Since its inception in 1975, PPS has worked with over three thousand communities in all fifty states on such projects. Post–Great Recession, the group has partnered with city governments on a regular basis: between 2008 and 2016, it worked on projects with municipal governments in such disparate cities as New York, Middletown, Connecticut, Richmond, Virginia, Corpus Christi, Texas, and Detroit. And companies have jumped on board as well. In 2014, PPS announced a partnership with Southwest Airlines. The "Heart of the Community" grantmaking program seeks to "raise awareness of Placemaking as a mainstream approach and a catalyst for building sustainable, healthy, inclusive, and economically viable communities." By 2015, the program's work in such cities as San Antonio, Milwaukee, Providence, and Baltimore was being feted in such publications as the US Conference of Mayors' "Mayors and Businesses Driving Economic Growth" annual report. Placemaking had spread to the corporate world.[35]

Here Comes the Neighborhood: The Return of a Concept

In 2007, *The Great Neighborhood Book: A Do-It-Yourself Guide to Placemaking* was published as "a Project for Public Spaces book." Authored by Jay Walljasper, it proceeded from the assumption: "The neighborhood is the basic unit of human civilization. . . . [T]he neighborhood is easily recognizable as a real place. It's the spot on earth we call home." For Walljasper, the neighborhood was a physical manifestation of "the level of social organization at which people interact most regularly and naturally." Not surprisingly, PPS and other groups engaged in placemaking efforts soon focused their attention on this important spatial unit.[36]

The timing of such a fixation was anything but coincidental. Throughout the first decade of the twenty-first century, a spate of sociologists cast great attention on the concept of neighborhood—in ways that would come to inform how placemaking was both thought about and practiced. Three of the most influential—Robert J. Sampson, Terry Nichols Clark, and Richard Lloyd—knew each other at the University of Chicago.

The university had been the site of the genesis of the influential Chicago school of sociology, which argued for the primacy of place-based communities in the century's rapidly expanding cities, particularly in its magnum opus, *The City* (1925).[37] Along with its European predecessors and counterparts,[38] the group identified the rapid migration (and immigration) of the rural population to urban centers as a defining characteristic of the modern era. Compelled by this abrupt transformation, members of the Chicago school argued what placemakers would argue nearly a century later: self-interested individual competition undercut previous social relationships in the city, substituting "organization based on occupation and vocational interest" for "family ties, local associations, [based] on culture, caste and status."[39] As globalization, urbanization, and capitalism spread, the city was where the old ways of relating to people were undone and replaced with more fragmented connections.

But this research expressed more than how markets influenced occupations and identity in the twentieth-century city. Chicago's Department of Sociology was actively studying the social forces of community life, including its physical characteristics, self-awareness as a community, and sorting of individuals according to occupation, nationality, and class.[40] Here, variations in geography emerged in finer detail as a potential expression of social distances or as potential conflicts with politically defined boundaries. The resulting neighborhood studies,[41] along with programmatic texts like *The City*, formed the concrete basis of the Chicago school and its influence on the conception of space. In the process, many studies tested whether the neighborhood—"a small, homogeneous geographic section of the city, rather than [a community,] a self-sufficing, co-operative and self-conscious group of the population"[42]—was losing its function as processes of competition reworked social life. Some scholars, like Ernest Burgess, were "not so certain."[43]

The sociologist Terry Clark joined the department in 1966 while the second wave of the Chicago school was hitting its stride, with texts like Morris Janowitz's *The Community Press in an Urban Setting* and Gerald Suttles's *Social Order of the Slum* demonstrating the limitations of the early ecological model's conception of social and physical connections while affirming the role of place in a revised model of social ecology.[44] This distinction was perhaps clearest in Janowitz's notion of communities of "limited liability," which sat conceptually between the school's early expectation of community decline and the stable friend and family networks of smaller communities. People developed these communities of limited liability when, for example, they chose to do things together. Among the most important of these places were schools, which focused family life on neighbors through numerous shared activities and their corresponding commitments.[45] The value of these connections was then reinforced through expressions of community solidarity in local newspapers and other symbolic forms of urban connection.

But Clark's own work was more firmly on the political side of sociology, connecting community to culture in a changing political economy. Among the first to describe what has come to be known as *third way* politics or the *new political culture*, Clark has published extensively on how the social shifts that have moved politics away from traditional party machinations have oriented local governments toward addressing social issues and lifestyle choices.[46] Cities may still compete for projects like the second Amazon headquarters, but they do it through selling the best mix of amenities (and, of course, tax breaks).

Clark's recent related research, with Daniel Silver and others, has been on *scenes*, social communities with their own aesthetic properties, judgments, and behaviors, often with connections to place.[47] Here, they found connections between Haight-Ashbury's hippies, Nashville's country musicians, and Chicago's gay neighborhoods. To Clark and his coauthors, these places—indeed, all places—were socially constructed by the people who connected with them, gave them identity, and created something distinctive around them, that is, an easily identifiable scene.

One of Clark's students, Richard Lloyd, has become one of the most referenced intellectuals in placemaking discussions and endeavors. His research on art and neighborhood development, published in 2006 as *Neo-Bohemia: Art and Commerce in the Postindustrial City*, explored Chicago's once-edgy Wicker Park neighborhood and the resident artists who valorized the very authenticity of the neighborhood they transformed in their quest for a neo-bohemian lifestyle. To Lloyd, the neighborhood famously chronicled in the film version of *High Fidelity* was not only a playground for consumption but also emblematic of how cultural innovation was intertwined with capital accumulation. Neo-bohemia was gentrification, but it was gentrification on the terms of the post-Fordist city, where employees of information and creative industries consumed and produced the culture of neo-bohemia in a landscape previously defined by manufacturing and its laborers. This dual consumption and production created a place where not only artists desired to be but also young professionals wanted to consume.[48]

If it sounds like Clark's and Lloyd's works may have informed Florida's argument about the creative class, it is not an accident. Both had connections with Florida that predated the publishing of *The Rise of the Creative Class* and are cited in that book.[49] But Chicago's influence on placemaking ran deeper than this. While this direction

Overlooking Chicago's Wicker Park neighborhood and the Loop
from one of the neighborhood's rooftop bars.

is seen through Clark's and Lloyd's political economic lens, Robert J. Sampson's effect was through a more traditional Chicago sociological approach.[50] For more than thirty years, Sampson has extended the early Chicago school work on social organization by linking concepts like *social disorder* or violations of a community's understanding of how people should behave in a place with a variety of other outcomes, including influential work on social capital.[51]

In his recent *Great American City: Chicago and the Enduring Neighborhood Effect*, Sampson argued that many contemporary attempts to understand the city and urban life have overlooked the importance of the local neighborhood because they have emphasized the way social connections have been weakened in the contemporary city. Globalists welcomed this change because it freed people from restrictive local social norms, while communitarians cursed the fraying of social bonds. "Either way," Sampson noted, "the implication many . . . have taken away is that places—especially as instantiated in neighborhoods and community—are dead, impotent, declining, chaotic, irrelevant, or some combination thereof."[52]

Yet Sampson made the point that, even in an increasingly globally integrated world some fifty years after Jane Jacobs voiced a similar sentiment in *The Death and Life of Great American Cities*, the processes that truly affected our life chances were based in "the *interlocking social infrastructure* in the neighborhoods of American cities." Neighborhoods were not simply places where people chose to live their individual(istic) lives, nor were they places made irrelevant by social processes that seemed to flatten the world. As did Lloyd's understanding of neobohemia and Clark's scenes, Sampson's emphasized how the mechanisms of cultural and social reproduction remained rooted in place-based communities. People still connected with each other in place, creating friendships, defining what was important to their well-being, and solving problems that challenged that well-being.[53]

What was important about Sampson's approach to the city was his argument that narrow readings of place and the city needed to be resituated and reinformed. In *Great American City*, he sought to "draw together, in one analytic framework, the diversity of the contemporary city" as well as to "transgress the narrow confines of disciplinary fields and canonical variables, and embrace a more holistic and systemic approach that gives priority to general social mechanisms and processes." How did he plan to do this? By probing "the social, organizational, and spatial contexts that produce collective civic events." Such ideas provided an updated intellectual framework for the place-making movement, one that jettisoned past understandings of urbanism and privileged social relationships while connecting them to a physical space that they already saw as important to their efforts—the neighborhood.[54]

We Are All Makers Now: The Rise of Mainstream Placemaking

Outside Chicago, other academics emphasized the roles that arts and culture could play for neighborhoods and the health of a city's overall economy. This work proved similar to but much more rigorous than the writing produced by Florida and often had the backing of institutions like the Rockefeller Foundation. More importantly, it came to inform the efforts of such organizations as the NEA, which have proved to be instrumental in the rapid rise of placemaking in the twenty-first century and the mainstreaming of such ideas and efforts.

Integral early work done in this process was completed by Mark J. Stern and Susan C. Seifert of the Social Impact of the Arts Project (SIAP), a research group initiated in 1994 to "explore the impact of the arts and culture on community life" and housed in the University of Pennsylvania's School of Social Policy and Practice. More specifically, Stern and Seifert have focused on "the relationship of the arts to community change, with a particular interest in strategies for neighborhood revitalization, social inclusion, and community wellbeing." SIAP's work also highlights the fact that many of the concepts that inform contemporary placemaking have their roots in the preceding century.[55]

For Philadelphia, for example, the late twentieth century was marked by an exodus of both people and industry. Of a population of over 2 million in 1960, just over 1.5 million remained in 2000. At the same time, deindustrialization wreaked havoc on the city's manufacturing-based economy. In 1975, 31 percent of the Philadelphia metro area's jobs remained in manufacturing. That number had declined to 14 percent by 1990 and 6 percent by 2016. Manufacturing employment fell 32 percent from 1980 to 2000, leading to declines in both jobs and real earnings. Faced with such daunting statistics, SIAP sought to identify new engines for economic growth for a twenty-first-century postindustrial Philadelphia.[56]

By the late 1990s, Stern and Seifert were issuing SIAP working papers with titles like "Community Revitalization and the Arts in Philadelphia" (1998). This report found that "sections of the city with a strong arts presence had greater population growth and a more rapid decline in poverty during the 1980s" and that "this revitalization does not fit common notions of gentrification." Moreover, it found that "patterns of participation of community arts programs contribute to revitalization by breaking down social and economic barriers separating communities" and that "community arts programs are strategically located to serve as facilitators of community economic revitalization."[57]

Importantly, such findings explicitly challenged the conclusions of academics like Sharon Zukin and Neil Smith, who coupled the arts with gentrification in cities such as New York. Those critical of this relationship, like Zukin and Smith, "tell a story in which the role of arts in *economic* revitalization is antithetical to processes of *social* revitalization." Yet, for Stern and Seifert, the true problem was the pace of urban change. "The crux of gentrification," they found, "is the *rapid* displacement of a low-income population by higher-income residents or

uses." So "when the process of small-scale redevelopment is drawn out over a long period of time"—the approach that Stern and Seifert called for—"it leads not to the classic displacement of gentrification, but to the construction of a particular kind of urban community—a community that must face up to diversity."[58]

Stern and Seifert also took issue with scholars of community development who argued that "the internal strengthening of minority communities is a critical strategy for rebuilding our cities." While not arguing against the strengthening of such communities, they found that this strategy has served only to isolate such communities. "Poor neighborhoods lack resources," they wrote. "They need employment, money capital, and integration into the wider urban community. Only strategies that better connect poor neighborhoods to the rest of the city are likely to succeed." Importantly, they did not see traditional neighborhood organizations—civic associations, neighborhood-improvement groups, and block clubs—doing this. But they did see arts groups starting to: "So, the connections fostered by the regional character of community arts form a new foundation for the social revitalization of the city's neighborhoods."[59]

And the findings seemed to back up such an externally focused approach. Looking at community arts groups throughout Philadelphia, Stern and Seifert noted that "the participant base in these neighborhoods comes disproportionally from *outside* of the neighborhoods in which the groups are located." As they admitted, at first blush this might be considered problematic: "We expect community arts groups to be based in their neighborhood. Yet, on further examination, this pattern can be seen as an asset."[60]

By late 2000, Stern and Seifert were still unpacking the topic of cultural participation and communities in an SIAP working paper from October of that year. But the subtitle of that paper—"The Role of Individual and Neighborhood Effects"—illustrated how their thinking was continuing to evolve. As scholars and others attempted to explain "the significance of the arts and culture in American society" the default unit of analysis had been the individual. In a nation so committed to the concept of individualism, this was not that surprising. Stemming from such a reality, the literature on the economic impact of the arts "has viewed culture as a set of individual consumption decisions around participation." In the end: "The total impact of the arts is simply the sum of many individual impacts."[61]

But, for Stern and Seifert, such an emphasis missed the social networks within which these individuals existed. Drawing from the work of Robert J. Sampson and Robert Putnam, they argued that those interested in gauging the economic importance of arts and culture had to look at broader communities. They built on Putnam's conclusion that "social networks are the critical mechanism through which *social capital* is developed," and they extended Sampson's idea of collective efficacy, or what they described as "a process through which geographic neighborhoods are transformed through the development of social networks." For Sampson, such efficacy was "the critical element in understanding a variety of child outcomes from physical health to cognitive development." And, in their examination of four American cities (Atlanta, Chicago, Philadelphia, and San Francisco), they concluded that "neighborhood effects are as strong as individual level variables in influencing the frequency of cultural participation." What did their findings suggest? "Cultural participation needs to be seen as a form of collective behavior. The over-reliance on individual level models of cultural engagement misses the very strong influence that social context has on participation." In the process, the ways individuals can reach better outcomes when rooted in a broader, supportive community are overlooked.[62]

Stern and Seifert reached similar conclusions in October 2005's "The Dynamics of Cultural Participation: Metropolitan Philadelphia, 1996–2004." Paying attention to the social context of cultural participation, they found

that higher-income communities seemed to have more of such activity. However, this may be because the cultural activities of those communities tended to be larger in scale and more formal and therefore more likely to be noticed. While cultural participation in lower-income communities needed to be encouraged, what was also necessary was a method of analysis that took into account "trends in cultural participation in small and informal cultural settings" as the limited data available suggest that such activities "represent a growing share of total cultural participation." Such activities had to be supported—and also documented—as part of a broader push for cultural equality. These ideas would inform placemaking campaigns in the years that followed.[63]

In 2007, SIAP initiated a Rockefeller Foundation–funded partnership with the Reinvestment Fund (TRF), a group working to inspire economic activity in low-income communities, on the topic of creativity and change. As part of this effort, Stern and Seifert authored a 2007 report on cultural districts that saw little good in building "big-ticket downtown cultural districts." Instead, the key to a successful "urban revival" lay in "using culture to revitalize the urban grass-roots, its neighborhoods, and their residents' civic engagement."[64]

This brand of urban culture moved away from such standard forms as the symphony, the musical, and the ballet and toward something "more active, more accessible, and more polyglot." In the early twenty-first century, artists were now "social entrepreneurs, selling their wares as well as their vision." So, while there was little doubt that "the arts *are* commerce," they fulfilled such a financial role primarily by building the ties between neighbor and community. "It is these social networks," Stern and Seifert found, "that translate cultural vitality into economic dynamism."[65]

This process of transforming cultural vitality into economic dynamism had the effect of encouraging innovation and creativity within the surrounding community. Soon, "geographically-defined networks created by the presence of a density of cultural assets in particular neighborhoods" would emerge; Stern and Seifert called these *natural* cultural districts. Unlike the big-ticket cultural redevelopment projects of the past—which often forced facilities into unwilling communities—this approach allowed spaces to develop more organically. As more and more of these "cultural assets" cluster together, the neighborhood gets closer and closer to "a regeneration tipping-point" and, in that process, continues "attracting new services and residents."[66]

A second report from the TRF-SIAP partnership, Jeremy Nowak's "Creativity and Neighborhood Development: Strategies for Community Investment" (2007), further explored the relationships between neighborhood, urban redevelopment, and creativity. On the very first page, Nowak, a banker, wrote: "I define community development as place-making." He saw placemaking as a collective endeavor, one involving households, businesses, government agencies, and civic institutions. First and foremost, these actors employed placemaking for practical reasons as such efforts could "increase economic opportunity, the quality of public amenities, and flows of capital into the built environment." Yet it was also an intellectual exercise, one that "manages a range of practical tensions: between market and civic capacities and roles; physical design and social utility; and the need to integrate the old and the new."[67]

In fact, with his assertion that "community-based arts and cultural activity" were central to placemaking, Nowak cast artists engaging in placemaking as actors able to reconcile the tensions between the individual and the communal within American cities. Such artists "are natural place-makers who assume—in the course of making a living—a range of civic and entrepreneurial roles that require both collaboration and self-reliance." The arts provide avenues for individual expression but, in the process, "facilitate connection across urban and regional boundaries"—connections that "reinforce and build social capital." And this social capital was, according to Nowak,

"not simply a by-product of prosperity but a potential precursor to prosperity."[68]

This accumulation of social capital through placemaking could have spatial as well as economic implications for cities. In a chapter titled "Architecture of Community: A Framework for Place-Making," Nowak, like Florida and Stern and Seifert, found that arts and culture had a "measurable impact on employment, investment and consumer spending." Such economic activity attracted new forms of arts and culture, which in turn attracted more members of the creative sector. Arts and culture thus "contribute to the qualities of place that, in turn, can attract residents, consumers and businesses." For Nowak, such processes helped cities redefine themselves for the twenty-first century. Placemaking endeavors could therefore have "significant implications for cities struggling to define post-industrial relevance."[69]

The NEA and Placemaking: From Our Town to Creative Placemaking

Yet perhaps the most important organization in the twenty-first-century rise of creative placemaking has been the NEA. On his appointment as chair in 2009, Rocco Landesman used Nowak's paper as the basis for his belief that "the federal government could be a more effective partner for community-based arts efforts." Nowak's work allowed him to see, he noted, "how the arts can be a force for social cohesion and economic development in neighborhoods, communities, [and] cities."[70]

Such a development fit in nicely with the intellectual trends of the first decade of the twenty-first century. But it also gave new direction to an organization that had appeared to be lacking direction for the previous twenty years. The NEA was a prime target of conservatives during the culture wars of the early 1990s. This moment led to a crisis in public arts funding as Republican-controlled government bodies worked to eliminate almost all federal funding for artists; many state arts boards, private and corporate philanthropies, and other donors made similar cuts. One end result of such cuts was that, in November 1994, the NEA eliminated the system that allowed local nonprofit organizations to earmark $1.3 million in federal grants for individual artists. This was a direct response to conservative backlash, particularly within a Newt Gingrich–led Congress, that took great offense at putting government dollars into the hands of such artists as Ron Athey, Barbara de Genevieve, Merry Alpern, and Andres Serrano. The NEA spent much of the rest of the century fighting off such attacks, leaving it little time to develop any sort of coherent agenda. A 2009 history of the NEA titled a chapter on the organization during the 1990s "What Is to Be Done?" The Obama administration would start to act.[71]

The concept of placemaking—and its relationship to economic growth—provided much-needed direction for the NEA. Commenting on this development, the economist and placemaking consultant Ann Markusen has found that "advocates for the arts began to shift their lobbying and research" away from looking for ways to fund individual artists and "towards underscoring the economic contributions of cultural industries and arts organizations." Within the NEA, this evolution came to a head under the leadership of Landesman, who served as chairperson until 2012 and pushed for the organization to adopt "Art Works" as its primary motto.[72]

The timing of Landesman's arrival at the NEA was opportune, the country still being mired in economic recession. Landesman explained: "I knew that if I went around saying, 'The XYZ arts organization is going to go out of business unless you give them some more money,' folks would shrug their shoulders and say, 'That's too bad, but we are dealing with a lot of problems, so good luck.'" Instead, he saw the recession as an opportunity to push his organization in a new direction, one rooted in the idea of what the arts could bring to a region rather than why the arts needed to exist. And creative placemaking became

the tool to make this case, particularly following the Great Recession. Creative-placemaking endeavors highlighted the cost-effective "ways that the arts change the social, physical, and economic characters of places." For Landesman, the goal was that any effort to jump-start economic growth and urban revitalization had "to include the arts": "Artists are great placemakers. They are entrepreneurs and they should be the centerpiece of every town's strategy for the future."[73]

Such ideas and language were highly informed by a 2010 NEA-commissioned white paper by Ann Markusen and Anne Gadwa Nicodemus that has proven to be perhaps the most influential document for the continued evolution of placemaking. For NEA chairperson Jane Chu and NEA director of Design and Placemaking Programs Jason Schupbach, the report cogently synthesized previous work as it "placed [Jeremy] Nowak's research in a new framework and coined the term creative placemaking." For Markusen, director of the Arts Economy Initiative and the Project on Regional and Industrial Economics at the University of Minnesota's Humphrey School of Public Affairs, the report allowed her to extend on the work she had begun in 2002's *The Artistic Dividend*. This work cast, in her words, "artists as a part of the export base" of a region's economy while also working to "support other industries and make them more productive." Gadwa Nicodemus, a principal at Metris Arts Consulting, came from an arts-administration background.[74]

Drawing on case studies from around the United States—though the bulk of the research was done over the phone—the report attempted to flesh out what the concept entailed.[75] Markusen and Gadwa Nicodemus clearly saw the work of others as vital to laying the groundwork for the ascendancy of creative placemaking: their document liberally cited the work of those who came before them. Very quickly, they settled on a definition of the concept: "In creative placemaking, partners from public, private, non-profit, and community sectors strategically shape the physical and social character of a neighborhood, town, city, or region around arts and cultural activities. Creative placemaking animates public and private spaces, rejuvenates structures and streetscapes, improves local business viability and public safety, and brings diverse people together to celebrate, inspire, and be inspired."[76]

Despite such an all-encompassing definition, Markusen and Gadwa Nicodemus honed in on the economic impact of creative placemaking. Such "creative locales," they found, "foster entrepreneurs and cultural industries that generate jobs and income, spin off new products and services, and attract and retain unrelated businesses and skilled workers." At the same time, the artists who tended to lead creative-placemaking projects improved the safety of communities by serving as eyes on the street, which then further encouraged development. Such individuals, "often traversing the neighborhood at all hours, make the streets livelier and safer, as do patrons of cultural venues and well-designed streetscapes."[77]

Importantly, creative placemaking did not repeat the past mistakes of large-scale, government-led urban renewal projects. As Markusen and Gadwa Nicodemus noted: "Governments have committed billions to physical infrastructure and incentives to induce companies to stay, with mixed results." Faced with diminishing budgets, cities no longer had the financial capability to carry out such projects. But Markusen and Gadwa Nicodemus would not recommend them even if they did as such physical capital investments "have crowded out human capital investments that hold greater promise for regional development."[78]

With an emphasis on developing human capital, creative placemaking moved beyond earlier urban redevelopment campaigns that focused on "tearing down and replacing aging factories and housing with monolithic districts and structures." Instead, Markusen and Gadwa Nicodemus saw creative placemaking as emerging from the specific history explored in the first two chapters of this book. Their vision of twenty-first-century placemaking, for example, "invokes what Jane Jacobs celebrated in post–World War II Manhattan—a mosaic of distinctive neighborhoods, each with its cultural hallmarks, cuisines, festivals, and street life." And, like Jacobs's championing of cheap space, creative placemaking allowed for "vacant auto plants, warehouses, and hotels" to be "transformed into artist studios and housing, infusing creative and economic activity into their neighborhoods." Ultimately, such places served as "incubators" for arts and cultural enterprises as "cultural industries flourish in creative places": "This role of creative placemaking in hosting cultural industries is under-appreciated."[79]

Markusen and Gadwa Nicodemus's mentioning of vacant urban spaces was telling. Like Landesman, they found that creative placemaking could speak to the problems associated with the twenty-first-century city. In fact: "In this difficult Great Recession era, creative placemaking has paradoxically quickened." The recession highlighted the fact that an overreliance on finance and tech jobs could get an economy into trouble. Sounding like Richard Florida, Markusen and Gadwa Nicodemus called on cities to diversify their economies by fostering the "cultural industries," which in the United States "are undisputed world leaders and innovators, responsible for millions of good-paying jobs." Such jobs ran the gamut from television to social media, from advertising to the visual arts, and from design to video game development. And the individuals who hold them "form a highly educated and innovation-producing segment of the American workforce."[80]

And there were a lot of them. In 2005, an estimated two million Americans reported work in the arts as their major occupation. This included writers, dancers, designers, architects, musicians, and visual artists. Including all those within the cultural industries, the total is between 4.6 and 4.9 million jobs, more than 3.5 percent of the American workforce. These people produced things that made money. Looking at a small set of cultural industries—broadcasting, telecommunications, motion pictures, sound recording, the performing arts, printing, and publishing—Markusen and Gadwa Nicodemus found that these job categories generated $45 billion in export sales in 2008, "more than computer system design, electrical equipment, air transportation, financial services, and American agriculture industries." Placemaking would help grow such occupations.[81]

Finally, placemaking efforts could also help boost tourism. "International tourism," Markusen and Gadwa Nicodemus found, "strongly tied to arts and culture, is an especially important source of export earnings. Visitors to the US spend much of their time and money visiting unique and prestigious cultural sites and enjoying live performances." "A place," they continued, "without a distinctive cultural aura is much less apt to land on visitors' itineraries than those with such amenities." Creative placemaking helped create such an aura and, in the process, attract, retain, and even create more "creative workers and entrepreneurs."[82]

The titles of the case-study chapters highlighted what such placemaking endeavors relied on: "Community Developers Partner with Theaters" (Cleveland); "Artists, the Third Leg of the Cultural Stool" (San Jose); "After Autos … Artists" (Buffalo); "Art Shores Up the Walk of Fame" (Los Angeles); and "Mayors and Artists Spark a Renaissance" (Providence). This connection between art and economic development was seen in the example of San Jose, where city officials aspired "for its downtown to be Silicon Valley's City Center." How best to do this? Bring in the artists! In early 2008, the city's Office of Economic Development/Cultural Affairs launched a citywide Creative Entrepreneur Project (CEP) "to nurture artists and link them with the region's extraordinary technology community." CEP provided artist business training, professional development scholarships, a web-based resource guide, and commissions for artists on public transportation projects. Explaining how such relationships worked, Kerry Adams-Hapner, director of cultural affairs for San

Jose, noted: "As inventors and interpreters of artwork, artists are now celebrated as the backbone of the arts sector, but also as small businesses that make San Jose cool, attracting talent and in turn economic activity."[83]

The "Creative Placemaking" white paper had immediate policy implications. It helped shape the guidelines for the NEA's first creative-placemaking program: the 2010 Mayors' Institute on City Design Twenty-Fifth Anniversary Awards. These twenty-one awards, including those recognizing public art integrated into a railroad track overpass in Greensboro, North Carolina, and a redesign of a portion of the Downtown Phoenix Public Market, "confirmed a demand for a federal program supporting arts-based development in US communities." This demand was addressed by Our Town, an NEA program initiated by Rocco Landesman in 2011. Between that year and 2015, the program invested over $21 million in 256 creative-placemaking projects in all fifty states, the District of Columbia, and Puerto Rico (applicants could apply for up to $200,000, with an average grant size of $75,000). The Our Town program helped instill, according to Ann Markusen, "a place-based mentality" within the work of the NEA. As the United States struggled to recover from the Great Recession, the emphasis on such a strategy made sense as, as Anne Gadwa Nicodemus noted, mobility was at a standstill and "people were stuck in their places."[84]

The program was meant to address such economic malaise directly. "The NEA created Our Town," Chu noted, "as a catalytic investment tool." And at the heart of such a mission was a very specific understanding of what creative placemaking actually was. "The simplest way to define creative placemaking," concluded Schupbach, "is as a way to strategically engage the arts in economic development priorities." Our Town looked to partner with other government agencies and efforts, including the White House's Promise Zones, an initiative developed by the Obama administration to press localities to partner with business and community leaders to grow local economies. At the federal level, Housing and Urban Development (HUD) and the Department of Education embraced creative placemaking through their Choice and Promise neighborhood programs. At the state level, in 2013 the Connecticut Office of Art announced that it would be focusing on placemaking endeavors as it made its grants. Finally, cities as diverse as Los Angeles, Milwaukee, and San Jose came to sponsor placemaking offices and programs.[85]

Soon, foundations across the country were coming to see the transformative power of creative placemaking. In Detroit, the Kresge Foundation was, by 2013, according to CEO and president Rip Rapson, "helping the city creatively reimagine the arc of its aspirations" through the funding of creative-placemaking projects. These projects not only spoke to those already living in Detroit; they also attracted new residents to the city. Such people were drawn by the promise of "working unconstrained by bureaucracy to carve out unexpected uses in unexpected places; converting the public ruins of factories into studio and exhibition spaces; and the opportunity to experience community vibrancy, street life, and cultural identity."[86]

Other foundations followed suit. To Darren Walker, president of the Ford Foundation, the foundation was already doing work with the arts and culture as a means "to enliven and enrich the community experience" while also doing programming that spoke to economic development. "Creative placemaking," Walker found, "was a way to bring together two long-standing areas of work of the foundation." Commenting further on this strategy, Xavier de Souza Briggs, vice president for economic opportunity and assets for the Ford Foundation (and previously a professor of sociology and urban planning), found: "In the past, community development and the arts were in parallel play. There had not been a lot of interaction, even when both were happening in the same places." Even when the arts were included in discussions of economic development, community development was often left out. Creative placemaking offered the means of bringing all these disparate strands together.[87]

For Briggs, creative placemaking was also valuable because it spoke to a changing understanding of what made urban spaces attractive. Over a period of ten to twenty years, such spaces had "gone from a near obsession with the hardware of place—the physical systems—to a much deeper appreciation for the role of human capital, knowledge, and creativity." The Ford Foundation's "support of creative placemaking reflect[ed] this shift." Creative placemaking also spoke to the changing nature of community development. As this concept continued to evolve, it took on "elements of a social movement, with community members making claims on government and the wider society." Strapped for resources by the early twenty-first century, governments could no longer afford to answer such claims on their own. Creative placemaking provided a way to address those shortcomings.[88]

In 2009, the Ford Foundation partnered with the MetLife Foundation to launch the Space for Change: Building Communities through Innovative Arts Spaces program. Funding for the inaugural year went toward the creation of a literary arts center in Minneapolis, the founding of a residency program for artists in exile in Pittsburgh, and the introduction of sustainable building and landscaping in the Watts neighborhood of Los Angeles, among other projects. The program's associate project director, Susan Silberberg, found that such projects highlighted how the shortcomings of traditional urban planning could be overcome. With such an approach, "the people become planners" as these projects "use placemaking as a tool to create the plan and to gauge the success of community engagement."[89]

Silberberg further articulated her ideas on placemaking in "Places in the Making: How Placemaking Builds Places and Communities," a 2013 report funded by Southwest Airlines. Like Markusen and Gadwa Nicodemus's paper, this document has come to play a profound role in the development of the concept. In a section titled "The Past and Present of Placemaking," Silberberg found: "Beginning in the 1960s, many of the current-day movements in city planning began to take root." Present-day placemaking efforts were an extension of these movements as they were also "a response to the systematic destruction of human-friendly and community-centric spaces of the early 20th century." And, while placemaking "has had many goals over time," "at its core it has always advocated a return of public space to people."[90]

For Silberg, the practice of placemaking had by the early twenty-first century come to concern "the deliberate shaping of an environment to facilitate social interaction and improve a community's quality of life." Yet what mattered here were the values promoted within individuals by such projects—and then the collective impact of such acts of promotion. The most successful placemaking projects thus privileged the intellectual over the material or the process over the product as they "transcend the 'place' to forefront the 'making.'" As this happened, placemaking led to the "empowerment of community through the 'making' process." Yes, the physical space of the site may change, but "the important transformation happens in the minds of the participants, not simply in the space itself." And this could then have an impact on broader communities. Placemaking can, as Silberg found, ultimately "address the pressing needs of our cities in a way that transcends physical space and empowers communities to address these challenges on an ongoing basis."[91]

On the one hand, this concept of transcendence highlighted that local physical conditions may not matter so much in developing a creative-placemaking strategy. Instead, it is the approach to making place—and the process behind such an approach—that should be privileged. Yet Silberg's work also suggested that older approaches to urban redevelopment needed to be transcended as traditional actors (government officials and urban planners) and conventional methods (urban renewal and other forms of large-scale urban development) no longer seemed able to speak to these pressing needs of American cities. Yet, perhaps more importantly, the practice of placemaking allowed the individual to transcend the outdated ideas of an earlier urban moment, the intellectual raw material that informed these earlier actors and development strategies. "The model of placemaking," Silberg found, "emphasizes flexibility, embraces impermanence, shares information, and draws on unortho-

dox sources for influence." As jobs, funding, and people fled cities, placemaking provided individuals the means to remake themselves—and, in the process, their immediate surroundings—for the postindustrial era. "We have gone from consuming places to making them," concluded Silberg concluded, an evolution that has "blurred the lines between layperson and professional—creating a community of makers."[92]

Like Markusen and Gadwa Nicodemus, Silberg provided ample case studies of sites she thought were getting it right, including the ubiquitous Bryant Park. But she also included sites from the Rust Belt, including Detroit's Eastern Market and the City Repair project in Cleveland. In such cities, the legacies of deindustrialization and disinvestment could be addressed in new ways; places that saw the loss of thousands of manufacturing jobs could once again empower residents to become makers. Eastern Market was therefore more than "a remarkably successful attempt to address food access while building community in a dramatically shrinking city." It was also a site of intense entrepreneurial activity: by 2013, there were over two hundred traditional produce vendors and fifty specialty food vendors, all working "to keep the customer base broad." In Cleveland, City Repair allowed community members to move beyond methods of redevelopment associated with the twentieth century and "to vision and collectively implement the type of shared space they want, without the help of official approval or institutional support." Again, the emphasis was on doing: the project built urban farms, planted community gardens, and installed murals. Yet, here, the payoff was again more than economic. Such work strengthened the necessary social bonds that produce a thriving city. Each City Repair project was initiated through community listening sessions and workshops to see what residents wanted to do. According to one project participant, such an approach provided more opportunity "to engage the neighborhood and create a stronger sense of attachment between neighbors." Placemaking, in other words, produced many useful assets.[93]

A Saturday morning at Detroit's Eastern Market.

It All Falls into Place: The Rise of ArtPlace America

Many of these individuals, ideas, and institutions played roles in the creation of the organization that, by the second decade of the twenty-first century, had come to dominate the world of creative placemaking. Beginning in 2011, ArtPlace America, a consortium of fifteen foundations, eight government agencies, and six financial institutions, used its National Creative Placemaking Fund to bring greater funding and visibility to placemaking. Partners included the Ford Foundation, the Rockefeller Foundation, the Kresge Foundation, the Knight Foundation, the Mellon Foundation, and Bloomberg Philanthropies as well as banks with a strong presence in community development, including Deutsche Bank, Bank of America, Chase, and Citi Bank. Other corporate partners included MetLife and Morgan Stanley. Government partners included the NEA, HUD, the USDA, the US Department of Education, the US Department of Transportation, and the US Department of Health and Human Services. Brought together by Landesman in a series of high-level meetings with foundation presidents and Obama administration officials following the release of the Markusen and Gadwa Nicodemus paper, the diverse partners agreed to participate in ArtPlace America for a ten-year period. Between

2011 and late 2014, ArtPlace America invested $56.8 million in 189 creative-placemaking projects in 122 communities across the United States. By January 2017, it had invested $96 million in 226 projects.[94]

There is little doubt that the work outlined above led to the creation of ArtPlace America. To Jamie Bennett, executive director of ArtPlace America, the organization embodied the idea of "arts and culture working to help achieve a place-based change." Like many who came before him, Bennett privileged both the creativity that he saw inherent in such work and the ways such creativity inspired various types of activity. "In creative placemaking," he noted, "'creative' is an adverb describing the making, not an adjective describing the place." Not surprisingly, this emphasis on making led the organization to see creative placemaking as a viable means of development. "ArtPlace," Bennett continued, "has adopted the language of community planning and development as the framework and context for understanding the impact of our investments." What this really meant was that placemaking could bring financial benefits to a community. In other words: "Creative placemaking supports economic diversity and place-based prosperity in the community, creating more opportunity for all."[95]

There was, of course, a social side to all this. Creative placemaking "ameliorates structural design problems" and "encourages beautification, engagement, and reimagining use" as it "connects people with opportunities and one another." Spaces were created where people want to be through placemaking—and this process created new relationships. Synthesizing work on the subject from Jane Jacobs, through Richard Florida, to Markusen and Gadwa Nicodemus, Bennett found that creative placemaking could help sustain that vital type of community—the neighborhood—"by clustering together different types of arts spaces along underused streets," thereby created "consistent patterns of foot traffic" that did more than "improve public safety"; they also worked "to drive a neighborhood's economy."[96]

Indicative of the type of placemaking projects championed by ArtPlace America was the West Broadway Arts Initiative, a project in Minneapolis, that the organization awarded a $250,000 grant in 2015. Through their parent organization, the West Broadway Business and Area Coalition (WBC), the West Broadway Arts Initiative used the ArtPlace funding to install over forty pieces of public infrastructure on West Broadway, including bike racks, benches, planters, and information kiosks—all designed by artists from the community. Such efforts would, according to the WBC, build on the successes of such programming as the FLOW Northside Arts Crawl—a three-day event organized by WBC that featured over three hundred performing and visual artists at close to thirty-five galleries, studios, theaters, and commercial spaces—and formalize a burgeoning local arts-based economy that had been growing informally, and somewhat chaotically, in the West Broadway Corridor. The ArtPlace funding would, according to WBC, "create a more vibrant, unique, clean, green, safe, and welcoming corridor for North Minneapolis residents and visitors."[97]

Dudley Voigt, the artistic director of FLOW, seconded such assessments. Even beyond the renovations to the corridor itself, he was most excited about the creation of Freedom Square, a new public space to be developed through ArtPlace funding on a previously vacant city-owned lot. First and foremost, the square would provide economic opportunities for artists and others who would benefit from having a space to bring people together, including yoga teachers, musicians, and chefs. Yet, like Jamie Bennett, Voigt thought that this strategy to increase economic development would also bring a new sense of security to the community. She noted: "We also see it as an amazing crime deterrent, activating otherwise inactive spaces."[98]

On the national level, others within the ArtPlace America consortium came to voice similar sentiments. In fact, Rip Rapson thought that, by the early twenty-first century, creative placemaking was perhaps the best development

The Occupy DC encampment in Freedom Plaza.

option: "ArtPlace staked out the ground that when you're dealing with the host of seemingly intractable, wickedly different problems of our time, arts and culture simply has to be at the center of that conversation." He continued: "Cultural creativity may well be the driving force of community revitalization in the 21st century." On the one hand, there was a strong financial component to such strategies owing to "the potential that arts and culture have to drive economic development." Yet placemaking also led to community revitalization in the way such projects seemed to stoke "attachment to community and to drive the kind of long-term visioning that a community needs in order to remain vital and healthy." By the 2010s, it appeared that creative placemaking had fully arrived.[99]

At the same time, actors around the country were drawing from markedly different networks and histories to establish projects that engaged urban space. For example, in the same year that ArtPlace America was created, a group of activists inspired by Occupy Wall Street, the anti-inequality movement with its own place-claiming program, organized under the name Occupy the Hood. In an agenda presaging the Black Lives Matter movement, the Cleveland chapter of the group identified such problems as mass incarceration, police brutality, food deserts, and foreclosures as disproportionately affecting economically

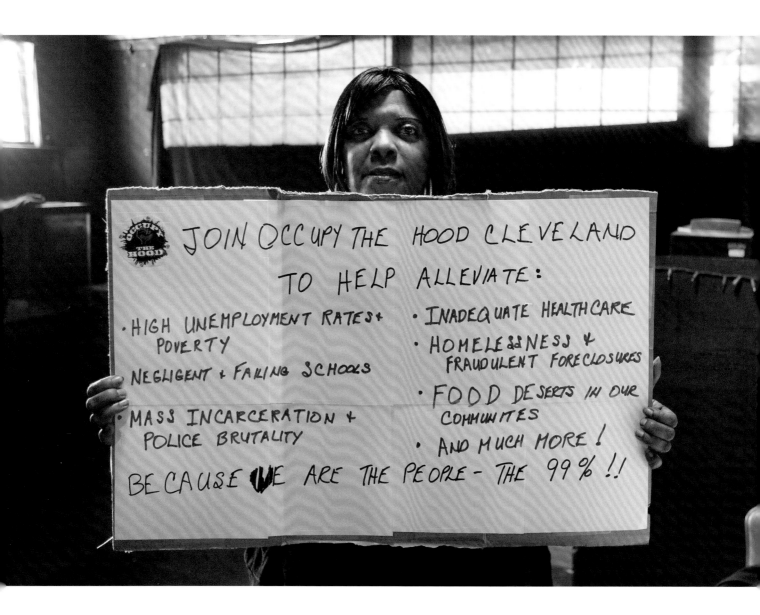

An Occupy the Hood activist during a community-awareness-
raising event hosted by the Cleveland chapter.

and racially segregated communities. In Milwaukee, the group demanded jobs to be relocated in the communities of color devastated by the closing of so many of the city's manufacturers. As Milwaukee's Kahlil Coleman told *In These Times*: "You look at cuts in jobs, schools, services, and it's mainly affecting the bottom of the bottom of the 99%." In order to understand and solve urban problems, place-based change would need to be understood as part of the problem and the solution to structural class- and race-based inequality. We now turn our attention to a history that illustrates how such counterapproaches and understandings emerged and how incorporating them into the intellectual underpinnings of creative placemaking is essential to its discourse and action.[100]

Growing Place

Toward a Counterhistory of Contemporary Placemaking

(*Previous spread*) A neighborhood resident harvests at the Sweet
Water Foundation's Perry Ave Commons.

In April 2017, Richard Florida did the unthinkable: he published a book that appeared to admit that he had got it wrong when it came to the relationships between cities, the creative class, and urban development. Well, perhaps *wrong* is too strong a word here. In *The New Urban Crisis: How Our Cities Are Increasing Inequality, Deepening Segregation, and Failing the Middle Class—and What We Can Do About It*, Florida conceded: "I had been overly optimistic to believe that cities and the creative class could, by themselves, bring forth a better and more inclusive kind of urbanism." Left to their own devices, those who made up the creative class simply moved to a small number of cities—and, within these urban centers, an even smaller number of neighborhoods. Such migrations ushered in a "new age of winner-take-all urbanism, in which the talented and advantaged cluster and colonize a small, select group of superstar cities, leaving everybody and everywhere else behind."[1]

What could be done to counter such movements? Curiously, Florida decided that, in the Trump era, the answers were to be found in government action, good old-fashioned urban planning, and public policy. "What we need is not a random menu of shovel-ready projects," he found, "but strategic investments in the kinds of infrastructure that can underpin more clustered and concentrated urban redevelopment. For infrastructure to really put the economy back on its feet, it must be part of a broader strategy for clustered, urbanized growth." For an academic who spent much of his career questioning the legitimacy of government presence in cities, such a position is startling.[2]

It is also quite costly. Florida realized that developing such infrastructure "at the scale that is needed [would] cost a good deal more than laying down roads and highways and throwing up single-family homes in the suburbs." At the same time, he also understood that, in the Age of Trump, "summoning up the political will to face up to the New Urban Crisis [would] be no easy thing." It is perhaps with such limitations in mind that he offered his only real policy: a tweaking of tax law. "The most effective approach to spurring denser and more clustered development," he concluded, "[would be] to switch from our current local reliance on the property tax to land value tax." Under such a tax model, urban property owners who build something like a parking lot would be taxed at a high rate, while small apartment buildings would be taxed at a lower rate, and larger apartment complexes would have an even lower rate. To Florida's mind, such a "system would provide greater incentives to put land in high-priced urban centers to its most efficient and productive use, increasing density and clustering." Needless to say, this approach is unlikely to be universally adopted anytime soon.[3]

Others had begun to question Florida's ideas, their underpinnings, and his acolytes well before 2017. To the cultural critic/historian Thomas Frank, such Floridian buzzwords as *creativity* and *vibrancy* obscured more than they revealed when it came to understanding the plight of American cities. More specifically, Frank trained his ire on groups like ArtPlace America that suggested—with little evidence—that "art production is supposed to be linked, through the black box of 'vibrancy,' to prosperity itself." If, indeed, "art creates vibrancy and increases economy opportunity," as suggested by one of ArtPlace America's slogans, what about those left behind? As Frank noted: "[Such organizations] have no comment to make, really, about the depopulation of the countryside or the deindustrialization of the Midwest. . . . They have no idea what to do for places or people that aren't already successful or that have no prospects of ever becoming cool."[4]

For Frank, the work of individuals such as Richard Florida and groups like ArtPlace America was not actually accomplishing anything, particularly in areas of the United States hit hard by deindustrialization. Others, such as Ian David Moss, have echoed these sentiments. In a widely circulated 2012 essay titled "Creative Placemaking Has an Outcomes Problem," Moss called out the ambiguous nature of ArtPlace's "vibrancy indicators," meaning the group's less-than-scientific tools meant to measure the complicated relationships between creative-placemaking efforts and urban economic growth. As he concludes: "Despite all of the new money that has been committed to the cause, creative placemaking still has an outcomes problem." What, in other words, does placemaking ultimately produce? And who benefits from it?[5]

Yet Moss was certain that it was the ideas of Richard Florida that informed these histories. He reminds us that it was Carol Coletta, then director of Smart City Consulting, who brought Florida to a national stage when she invited him to serve as coorganizer of the 2003 Memphis Manifesto Summit. Here, a wide audience of entrepreneurs, cultural leaders, and policymakers from around the country were introduced to his concept of the power of the creative class. Coletta herself would bring his ideas to the federal government when she became the inaugural director of ArtPlace America in 2011. As we have already seen, Florida's work continued to inform the evolution of placemaking well past 2011.[6]

If the Floridian turn has not led to a comprehensive matrix of placemaking, where else can one turn? One obvious answer is Jane Jacobs. As the shortcomings of Florida's work have become more apparent, the past several years have seen a spate of books, articles, and films calling for a reassessment of Jacobs's work and importance. The least-useful of such works, including the 2016 documentary *Citizen Jane: Battle for the City*, do little more than present Jacobs as caricature: the feisty community activist with eyes on the street. At the same time, more careful analysis suggests that she is somewhat implicated in the current moment. As we have already seen, she plays a vital role in how the field of creative placemaking has evolved and been legitimized. Yet Jacobs was writing before the racial upheaval of the late 1960s and at a time of less conspicuous economic inequality in many US cities. In other words, she may not be the best guide when navigating issues like race in the twenty-first-century city.[7]

But what about class? As the historian Brian Tochterman has noted, Jacobs argued that "much of what makes cities so attractive was already there—communities must nurture their own neighborhoods and the state should remain relatively hands-off." Such a market-driven approach did little to ensure a continued place for poor and working-class people in neighborhoods like Jacobs's West Village community. As the architectural historian Max Page concludes—in the introduction to a collection of essays titled *Reconsidering Jane Jacobs*—Jacobs was ultimately successful in saving her own neighborhood from the state-sanctioned forces of urban renewal. However: "She did not foresee what it might become—an area nearly uniformly for the wealthy." The concept that cities already had what they needed and simply needed to preserve it also did not speak to cities already ravaged by the worst of urban renewal or beginning to suffer from deindustrialization. Jacobs works when you have people, but what about cities suffering from intense population loss?[8]

In very real ways, these shortcomings of thinkers like Florida and Jacobs have come to inform the contemporary approach to creative placemaking. And they are tied to very specific histories of the mainstream creative-placemaking movement and the ways these histories are not employed in creative-placemaking endeavors. Despite Florida's recent mea culpa, he, along with Jacobs, was always suspicious of government. And, in her later work, as did Florida throughout his writings, Jacobs took for granted the death of the industrial economy and the emergence of a postindustrial one. Indeed, both basically see such a history of declension as a way to rethink and renew American cities, with an emphasis on creative types and the creative economy in larger cities like New York and San Francisco. This does not mean that either theorist was entirely wrong. In fact, both—and particularly Jacobs—offer ideas that could serve as jumping-off points for a more robust, more inclusive understanding of placemaking. But how to begin this important process?

As we alluded to briefly in previous chapters, there are histories that mainstream theorists and practitioners alike neglect when constructing an intellectual lineage for their work. In this chapter, we take such overlooked narratives seriously and offer a case-study counterhistory rooted in the Rust Belt city of Milwaukee. This counterhistory is meant both to complement and to challenge versions of the past that inform mainstream contemporary creative placemaking. By recognizing these parallel histories in a place few observers examine, it is our hope that we can offer a new wellspring of ideas for placemaking efforts. Far from the sources of the creative economy, we see an urban history that sets the contours of a new approach to creative placemaking, one that deals with the realities of inequalities while highlighting a continued commitment to production, a close attention to civil rights, the importance of organized groups, the potential for a reinvigorated public sector, and a profound appreciation of the ways that alternative urban cultures have significant spatial impacts.

Focusing on Milwaukee lets us see how history informs placemaking in ways that many contemporary practitioners and observers have not yet fully realized. At the same time, it highlights a history that is often overlooked, shed in favor of a postindustrial urban order. Counter to contemporary neoliberal arguments, that is not the only choice. We want to highlight other options and the deep roots of those options. In this chapter, all these issues will coalesce around the budding field of urban agriculture, in which Milwaukee has emerged as a hub with global influence. To its practitioners, urban agriculture is more than about growing food; it is about growing community and new forms of urban redevelopment. It is also about creating and navigating new relationships between such disparate partners as community members, for- and not-for-profit entities, and municipal government. And this history is but one of many; there are others that do or could inform alternative approaches to contemporary placemaking efforts. We hope that others continue this process of highlighting these important stories, and we hope that this case study provides a model of how to do such work.

"It's making something out of nothing": An Alternative Approach to Creative Placemaking

So how best to make sense of such ideas? The African American artist, activist, and educator Fidel Verdin begins to provide a model in Milwaukee. Since 2003, Verdin has organized citywide "Summer of Peace" youth rallies. These events, staged at public parks across the city, use art and culture to address youth violence. By 2014, Verdin had grown tired of moving from park to park and established a permanent home for such efforts, founding the Peace Park and Garden in Harambee, an African American neighborhood on the North Side of the city. Since then, he has hosted Summer of Peace events there as well as yoga classes, art builds, outdoor markets, community meals, and food-preparation demonstrations.[9]

The centerpiece of the park is a large cube, eight feet by eight feet, displaying the work of local graffiti artists. As Verdin explains: "We use the attraction of graffiti, of making public art, to spark curiosity [regarding urban agriculture] in children and adults." Verdin also serves as codirector of TRUE Skool, a nonprofit that, according to the organization's mission statement, uses "Hip Hop Culture & Creative Arts as tools to engage youth in social justice leadership, education, entrepreneurship and workforce development." Not surprisingly, then, hip-hop informs not only the aesthetic of but also the philosophy behind Verdin's work. For the Peace Park and Garden site, Verdin went "back to that essence of how hip-hop was created out of young people sort of place making and creating their own way, their own place, to celebrate whatever they wanted to do." Interestingly, his work does seem illustrative of the benefits of the hands-off approach to urban redevelopment that Jacobs championed some fifty years earlier. The philosophy of Peace Park was, according to Verdin: "You bring your resources, I'll bring my talent. It was like the early hip-hop idea of 'You bring your speakers, I'll bring my turntable.' That's sort of the spirit—it's

Fidel Verdin weeds in his Peace Park.

making something out of nothing…. We're going to piece it together along the way. We're going to freestyle it." Yet this was not necessarily an ideological commitment; it was something far more practical. Much like the early days of hip-hop, absent any avenues of formal support Verdin had little choice but to go the do-it-yourself (DIY) route.[10]

Rooted in Verdin's appreciation of hip-hop is an understanding of how issues of race, civil rights, and history inform his efforts to remake the urban fabric. "You can't," Verdin implores, "take race out of it if you're saying 'urban.'

It's a Black city." At the same time, he sees his work as an extension of the civil rights struggle in which his parents participated in the 1960s—and the legacies of political radicalism that the historical moment generated (his parents named him after Fidel Castro). In fact, he sees in this earlier era a model that he explicitly seeks to emulate: "I think back to the '60s and the civil rights movement and that era, and there were some dynamic young people that were leading and creating this mass movement." He sees himself doing similar work in the twenty-first century, in a way that builds on earlier civil rights struggles and recognizes the importance of that history to contemporary projects: "We definitely are a continuation, and it's a constant thing where we're trying to figure out the best way to infuse all that history and information into all the work we're doing today." To highlight this, a second park that Verdin worked to create in the neighborhood—Martin Luther King Jr. Peace Place—features murals that tell the story of the Milwaukee Commandos, a group of African American civil rights activists who fought for open housing in the city in the 1960s.[11]

But the park is about more than recovering and noting history: Verdin wants to use place to rebuild community. Many African Americans in cities such as Milwaukee are, Verdin feels, "disconnected from our traditions, from our folkways": "Those are the things we did down South: the land was a healer; the land is therapy. Being able to use community spaces to do that is healing for everyone." It is with such an idea in mind that he has come to think of his parks "as almost like a retreat thing." In light of things like poverty, eviction, and police brutality, how can places address the trauma and violence within the city? Or as Verdin pointedly poses this question: "How do we escape right in the 'hood, right there, right next to the terror shop, the Fifth District [Milwaukee police headquarters, a block away from Verdin's Fifth and Locust site]. How do we create a safe space that people know about? The same way people call protests at certain spaces, you call a meeting at the Peace Park." Again, one detects a whiff of Jane Jacobs's disdain for government power, yet Verdin would come to direct this disdain in ways that Jacobs was simply unable to.[12]

Verdin understands that many would consider the work he has been doing for over fifteen years creative placemaking, and he knows that his work "gets thrown in there." "And that's OK because I understand how it fits in there," he explains. "I never really knew it was a thing until I was already doing what I was doing. It's kind of funny to discover that there's almost a standard way that this is done." For Verdin, placemaking is "a buzzword ... a fundraising tool." Does he shy away from applying the term to his work? "Not really. I guess now it's like a layman's term; you can say it, and people understand what you're talking about." But, in explaining the impetus for his involvement in this field, he returns to the hip-hop analogy: "We went into it really like freestyling, not really having a script or template or lingo for it. We really just thought, Let's go crazy with these spaces and see what happens."[13]

Such a mindset does not mean that Verdin ignored the efforts of those who came before him. As in knowing civil rights history, he knew that he was not the first to intervene in the city's built environment. He realizes: "There's been a lot of creative-placemaking efforts that have happened for years under the radar, unsanctioned by the city. And they still exist—that's a real thing. People still take over their neighborhood places, their vacant lots, and that happens all the time, and it's been happening for years." From 1973, when the Sherman Park Community Association took over land earmarked for freeway construction, to 2014, when the community activist Andre Lee Ellis began growing food on a vacant lot elsewhere on the North Side, Milwaukee has a rich history of community members reappropriating underutilized land. It is this lack of awareness of these efforts that makes Verdin hesitant to join the placemaking conversation. "The conversation on creative placemaking?" he asks. "Eh, I don't know." At the heart of this hesitancy is his displeasure with the idea that one can "make" a place already ripe with such histories: "[It] sounds a bit colonial to me." Instead, at the heart of his approach is an engagement with such histories through the act of actually growing things. At both his parks, a commitment to producing healthy fruits and vegetables can be seen in raised garden beds and surrounding fruit trees. The Peace Park produce goes directly to needy nearby residents. At Peace Place, produce is used for culinary training at the not-for-profit organization HeartLove Place. The program provides recently incarcerated men and women with the skills needed to reintegrate into the workforce.[14]

In 2015, Verdin enrolled in Growing Power's Commercial Urban Agriculture Training Program (CUA). This program, initiated by the Growing Power urban farm founder and CEO Will Allen, took place at the Growing Power facility, founded in 1993, and perhaps the most renowned urban farm in recent history. To Verdin, Allen is a "living legend" and a true placemaker, as his efforts have led to the training of over one thousand people and the creation of countless new urban agricultural sites around the world. He appreciates how Allen sees access to healthy food as a "civil right," one "every bit as important as access to clean air, clean water, or the right to vote." According to Verdin, Allen's "Good Food Revolution" updates the civil rights movement to address "environmental justice, food justice, access to water": "It raises awareness on all those things."[15]

Ultimately, Verdin's experiences with Growing Power led him to rethink the concepts of creative placemaking and urban development as practiced citywide. Urban agricultural endeavors, Verdin came to understand, brought new resource-producing sites to parts of the city commonly left out of traditional redevelopment plans and mainstream placemaking initiatives, all the while shedding light on the inequalities created by these other histories. "So then," he concludes, "you start adding things up: where are the resources at? And then, naturally, if you're talking about gardening, you're talking about food. And then there's a conversation about food deserts." Soon, one comes to look at the vacant lots that house these projects in a different way: "It's not just about the empty lots. Businesses are empty; houses are boarded up. And it's only happening in a designated area. It's not all over the city." Once people begin to see this, "they'll start trying to take these resources" into their own hands: "There's so much possibility in fusing the activism, if you will, the civil rights, the culture, the human rights—whatever you want to call

it—with urban ag." Whether it be in the form of Verdin giving space to hip-hop graffiti artists in his gardens or the African American urban farmer Venice Williams selling herbs grown in her Alice's Garden (which itself took root on once-vacant land on the North Side of Milwaukee), urban agriculture provides new ways to think about place in American cities.[16]

Finally, Verdin implicitly understands that urban agriculture provides the model for a bottom-up approach to placemaking. He noted: "My first mind frame was that we were going to go into it [urban agriculture] without cooperating with the city, without communicating on that level." Yet, primarily due to the work of Allen and those he was inspiring, the city was paying attention: in December 2010, for example, the city government passed zoning changes that allowed seasonal markets to operate up to 180 days a year and allowed for the raising of crops in commercial and institutional districts. The following year, the city relaxed restrictions on the construction of hoop houses used for agricultural purposes. Given such responses, Verdin now believes: "Getting the city officials actually involved from the beginning of planning our projects and helping with resources for the building process is the goal. We want the city invested in itself and to see our long-term vision." In fact, the city has worked with Verdin to find funding for his projects, partnering with him and his partners on a 2014 Partners for Places grant offered, through a collection of funders including Bloomberg Philanthropies, the John D. and Catherine T. MacArthur Foundation, and the Surdna Foundation, to construct the Martin Luther King Jr. Peace Place. Here, the Jacobsian approach taken by Verdin has created a new way for governments to lead from behind in urban redevelopment.[17]

And this is not just a Milwaukee story. Here, we use the city to shed light on the networks—historical and contemporary—that drive these ideas and projects across the country. In Milwaukee, one sees the way city govern-

Venice Williams cuts herbs in Alice's Garden.

ment, nonprofit institutions like universities, and charismatic leaders such as Fidel Verdin and Will Allen interact, providing a potential alternative model for placemaking endeavors elsewhere, even outside the realm of urban agriculture. Moreover, an emphasis on this subject is particularly useful because urban agriculture has a long self-awareness of the skills and materials needed to grow plants or create a working farm. Often, this awareness has led to a self-consciousness about urban agriculture's place in a broader history and in broader redevelopment narratives. This approach, or perspective, is valuable for any sort of alternative placemaking. In the proceeding chapters, we will see how it can work with projects outside urban agriculture.

There Grows the City: A Brief History of Urban Agriculture in Milwaukee

Just as our historical approach to the development of placemaking in the previous chapters demonstrates the shortcomings of the typical conceptualization of placemaking, Verdin's story suggests that urban agriculture is

an ideal case to expand it practically and theoretically. Moreover, from City Slicker Farms in Oakland, California, to Grown in Detroit, advocates around the United States position urban agriculture as a way to empower communities, provide important skills in a new economy, and locate change in places deeply meaningful to participants. These attributes and others are outgrowths of the origins of urban agriculture as an answer to a litany of urban problems, a context not often understood by placemakers. Of course, traditional placemakers would say that their work often accomplishes the same goals. Yet urban agriculture's emphasis on the production of tangible goods distinguishes it from other placemaking strategies.

Historians and laypersons alike overlook the fact that there is a long history behind the work that Verdin is doing. A closer examination of the history of urban agriculture in Milwaukee allows us to see this narrative of declension in a new light and may even allow us to challenge it. If Milwaukee's growth was centered on industrial manufacturing for much of its history, agriculture and the various food-related goods and services that went along with it also fueled development. There is little doubt that such production led to increased economic development in Milwaukee. Yet, perhaps more importantly, food also led to the development of nonmonetary outcomes, including stronger neighborhoods, ethnic solidarity and acculturation, and civic identity, potential core ingredients of a new understanding of creative placemaking.

As noted above, Milwaukee is not the only urban center with such a past: Buffalo, Detroit, Chicago, Minneapolis, Cleveland, St. Louis, and other cities share similar histories of agriculture and urban development. In part, we focus on Milwaukee because it is so illustrative of these often-overlooked histories; this is not a tale of simple exceptionalism, and such commonality, we believe, serves only to make our case stronger. Yet we also believe that actors in Milwaukee have proved particularly adept at utilizing this history for contemporary placemaking endeavors. And, perhaps more importantly, other efforts across the United States have looked to Milwaukee for both inspiration and guidance as similar work has been undertaken. As the twenty-first-century postindustrial city begins to develop spaces of production once again, the influence of Milwaukee looms large.

The histories that inform actors and efforts in Milwaukee also begin to illustrate the intersection of race, culture, place, and community that informs urban agricultural endeavors around the country. There is a contemporary effort, as the sociologist Monica M. White notes, "to connect contemporary urban farmer-activists to an earlier time when African Americans turned to agriculture as a strategy for building sustainable communities." Some of this was undoubtedly a rural story, as seen in the work of Fannie Lou Hamer and her Freedom Farm Cooperative, founded in Sunflower County, Mississippi. Yet White reminds us that the city-based Black Panther Party offered free breakfast and grocery programs and supported Black farmers as a way of tapping into "the rich legacy of growing food as a strategy of survival and resistance." This chapter will further mine this legacy, particularly as it played out in Milwaukee, in an attempt to offer a counterdefinition of placemaking for the early twenty-first century.[18]

Despite such legacies, it may seem odd to locate a history of urban agriculture in Milwaukee. After all, the city was known as the "Machine Shop of the World" throughout much of its history as firms such as Allis-Chalmers and Pawling and Harnischfeger turned out large-scale industrial equipment for customers across the globe. And, when this era came to a close, city and private-sector leaders alike struggled to embrace a service-oriented economy that would help address the traumas associated with the deindustrialization process. Yes, Milwaukee remained known as a city devoted to beer, that special industrial-agricultural hybrid, throughout its economic dislocations, but even this characteristic came undone. By the end of the twentieth century, such celebrated breweries as Pabst and Schlitz had shuttered their Milwaukee plants.[19]

(*Facing*) Grown in Detroit's produce stand in Eastern Market.

Children run through rubble near Milwaukee's former Pabst brewery.

Commenting on this region's relationship to nature, geography, and industry, the Milwaukee historian John Gurda has noted: "Southeastern Wisconsin grew up at the intersection of industry and agriculture." This was where "fertile soil, inviting harbors, and a young, enterprising citizenry" helped the region "develop economies that blended ambition and innovation with strong ties to the land." In the first generation of white colonization—from the pioneer land sales of the 1830s through the Civil War—the relationship between the lakeshore cities and their food-producing neighbors was a matter of transportation. Milwaukee shipped out the crops that farmers from such nearby counties as Waukesha, Washington, Ozaukee, and Walworth brought in. By the early 1860s, the city had become the largest shipper of wheat in the world, and the Grain Exchange, an imposing structure located in downtown Milwaukee, "influenced the price of farm commodities internationally."[20]

Yet agricultural endeavors also fueled development within the city limits. Throughout the nineteenth century, recent immigrants to Milwaukee—particularly those settling on the city's South Side—brought with them the experiences of farming in the Old World. Not surprisingly, such individuals took to working the soil in their new home as well.[21] As they did in other late-nineteenth century cities, farmers in Milwaukee realized that they needed ways to get their products into the hands—and mouths—of consumers. In the early 1880s, a petition was presented to the Milwaukee Common Council requesting that the city organize and operate a space previously known as Haymarket Square as a public market. By the early twentieth century, the renamed Central Market had become a vibrant market space.[22] Other markets quickly sprung up, collectively selling over 13,300 carloads of fresh fruits and vegetables, with an estimated value of $24,078,600, by 1946.[23]

Throughout the mid-twentieth century, in addition to the Central Municipal Market, Milwaukee featured four other public markets: the Fond du Lac Market, the Cen-

ter Street Market, the East North Avenue Market, and the National Avenue Market. Collectively, these markets moved large volumes of vegetables, fruit, poultry, eggs, and flowers each growing season. They functioned as spurs for increased economic development in the neighborhoods that they served and helped create a truly local economy of food products as restaurants and small grocery stores often set up shop near them.[24] Yet these sites were also perceived as places where nature tempered the excesses of the city. "These markets are pleasant places," observed one 1940 customer, "retaining even in the midst of the city's bustle an atmosphere of a more quiet and leisurely way of life." In fact, as they did in cities like Cleveland and Kansas City, the markets helped contribute to Milwaukee's identity as a city where green public spaces were to be valued, an identity initially cultivated by the city's early socialist politicians and planners. Here, one could see "a bit of the great outdoors in all of its green profusion transplanted to gray concrete."[25]

However, the Central Municipal Market shuttered its doors in 1967, the Fond du Lac Market closed in 1971, and the Center Street Market served its last customer in 1980.[26] For many observers, the shuttering of latter reflected the city's changing population base. The Center Street Market had been a crucial part of the European immigrant existence throughout its long history. Now that deindustrialization was kicking in and countless individuals were fleeing to the suburbs or dying off, there were no longer any customers to sustain such time-honored traditions. On the announcement of the market's impending closure, the *Milwaukee Journal* ran a piece with the title "The Center St. Crowd," cataloging recollections of old-timers—all white—on the market. One elderly women, Margaret Palmer, told the paper: "I've been coming here since I was a child."[27]

Yet, in June 1981, the city opened the Fondy Market approximately three-quarters of a mile from the recently deceased Center Street Market. This municipally run space attempted to modernize the market for the late twentieth

A Hmong vender sells flowers and vegetables in the Fondy Market.

century in a neighborhood that was now predominantly African American in composition. As the city's ethnic white population declined, Milwaukee's African American population increased, going from 105,088 in 1970 to 146,940 in 1980. The Fondy Market sought to capitalize on this growth and kick-start redevelopment of an area hit hard by white flight as the children of the Polish and German immigrants that Margaret Palmer remembered left for the suburbs. At the ribbon-cutting ceremony for the new facility, Mayor Henry Maier noted: "We are taking a giant step toward restoring the Fondy business area to its place of prominence."[28]

With the influx of African Americans moving from the rural South to the urban North, there was a genuine push from new Milwaukee alderpersons and residents to have this new market speak to the community's new residents. An October 1988 ad for the market, for example, highlighted the fact that it carried sweet corn, sweet potatoes, and okra, food products that historically played significant roles in the cooking and eating habits of African Americans. All this food was grown in the metropolitan region, some even within the city limits. Yet, even when Fondy vendors grew their crops on the city's periphery, they were drawing customers to the market, and those customers often bought things like prepared foods from other vendors who lived in Milwaukee. Because of its ability to speak to the economic imperatives of a changing Milwaukee, the Fondy Market remains an important city asset in the twenty-first century. In the summer of 2017, a creative-placemaking endeavor led to the creation of a new public park, Fondy Park, located next to the market.[29]

Agricultural Extension Agents Trade the Farm for a Vacant Lot

The relationship between agriculture and redevelopment of the urban landscape stretched, as we have seen, all the way back to the nineteenth century. Yet, as was the case in cities across the country, urban growing came to be a part of more city residents' lives during World War II. Under the leadership of the University of Wisconsin—Extension employee and chairman of the Milwaukee County Victory Garden Committee Alex Klose, the proportion of Milwaukee-area households growing their own vegetables, in gardens usually twenty feet by thirty feet, quickly rose from 30.2 percent in 1942 to 54.8 percent in 1943. Milwaukee County opened ninety acres of parkland to city dwellers without yards of their own, and institutions like the police department set up gardens at places like the precinct station at Forty-Seventh and Vliet.[30] Under Klose's direction, a remarkable 123,000 victory gardens were planted in the county annually. The *Milwaukee Journal* reported that, in 1944, the program was rated by the federal government as first in size of gardens, first in management of these gardens, and third in total number of gardens in a metropolitan area.[31]

By the late twentieth century, those calling for a return to urban agriculture would revisit the legacy of the World War II–era victory garden. As farming practices continued to evolve—and as deindustrialization produced wide swaths of vacant lots throughout the city—new generations of urban farmers came to reimagine the cityscape as a place of agricultural production. Some of these

growers sold their produce at markets like Fondy, while others started small businesses. Some grew for their block, while others grew strictly for their own personal consumption. In 1978, the Shoots 'n Roots program, co-sponsored by the University of Wisconsin—Extension, the City of Milwaukee, and the USDA, reached out to 3,300 participants who helped plant over 1,300 gardens at 29 citywide sites, with a total of 200,000 square feet under cultivation.[32]

Such narratives inform the post-1960s history of urban gardening in Milwaukee. Beginning in the 1960s, city officials partnered with a variety of institutions to ameliorate the challenging urban landscape that confronted Milwaukee's growing African American population, evidence of which is seen in the spate of market closings that occurred from the late 1960s through the early 1980s. Groups such as the Milwaukee Urban League, area 4-H clubs, and the University of Wisconsin system (and in particular the University of Wisconsin—Extension program), in collaboration with city, state, and federal officials, looked to urban gardening programs as a means of addressing such issues as rising unemployment, increasing poverty, and violence. As in the past, city officials and other concerned citizens believed that programming based on urban growing could teach city dwellers values and skills while providing them with a tangible, useful product—healthy food. At the same time, participants and observers alike came to see these endeavors as also providing alternative models of urban renewal and redevelopment for a city feeling the pains of deindustrialization, helping heal neighborhoods scarred by the realities of abandonment while working to build new communal bonds among neighborhood residents forced to confront these emerging landscapes of vacancy. Previously, as seen during the Great Depression and World War II, such urban growing programs predominantly reached out to the city's white population. Yet as the demographics of Milwaukee shifted, the focus of this programming shifted as well, and the city's African American population came to be the primary target of outreach efforts associated with urban gardening. In the process, this section of the city's population would come to redefine such projects based on their own strengths and interests. Such a trend would continue well into the twenty-first century.

One sees evidence of the roots of such developments as early as the fall of 1962. The minutes for a November 1962 Milwaukee Urban League meeting noted, for example, that the group had been working with Carl Smith, the Milwaukee County 4-H agent of the Milwaukee County Agriculture Extension Service, on urban gardening programs for the past two years. "The need was expressed," according to the meeting minutes, "for further development of the program under the direct leadership of the County 4-H Agent's office."[33] A Milwaukee Urban League memorandum from January 1963 communicated some of the efforts that came out of these discussions, including a garden centered at the home of Mrs. Nellie Thomas, at 534 West Chambers Street. Here, twelve children (eight boys and four girls) had been involved with a successful community garden project.[34] The success of such endeavors led Smith to write to the Urban League 4-H leaders in March 1963: "I have been real pleased with the progress of the 4-H Clubs in the Urban League area." He was particularly enthusiastic about the work being done with woodworking and food production.[35]

Smith also realized that it would make sense to reach out to educational institutions within the state to assist with these programs. By the mid-1960s, the University of Wisconsin system began looking for ways to start better addressing the growing African American population in cities like Milwaukee. More specifically, university administrators sought programming that spoke to the lives of urban residents *not* currently a part of the world of higher education. The authors of a September 1963 report on what such outreach efforts should look like noted that the University of Wisconsin system had to reach out to adult residents of cities like Milwaukee, finding: "Some universities have concentrated their activities on graduate and undergraduate instruction and research. The Uni-

versity of Wisconsin has long maintained that its role is larger than this."[36]

The University of Wisconsin system had a history of reaching beyond campus boundaries and out into the real world, taking a "national leadership role in agricultural extension." It was such a role that needed to be revisited in the 1960s—but in a different context. The authors of the 1963 report referred to the efforts of University of Wisconsin-Extension program, which had made a name for itself primarily in the worlds of agricultural and horticultural outreach. This method of engaging with broader, primarily rural communities could prove to be a useful model for efforts to reach out to urban residents of Wisconsin. Equally as important was the fact that, at that same moment, the Extension was also searching for ways to make itself relevant to such urban populations. The Extension program, in other words, wanted to get off the rural farm and into places like inner-city Milwaukee at the same time that the university system itself wanted to start better addressing what was coming to be termed the *urban crisis*. Not surprisingly, a variety of groups working in vulnerable neighborhoods in Milwaukee welcomed such a development.

By April 1964, the Extension had established the Urban Work Group, meant to address the question, "What ought Extension do in the coming year relevant to that package of urban problems conjured up the somewhat epithetic label, 'the inner core'?"[37] During the summer of that year, the Urban Committee had come together to attempt to answer such a query. As the committee saw it, efforts to address the plight of the urban poor needed to focus on jobs and labor. More specifically, committee members voiced their concern regarding the pool of predominantly African American unskilled laborers who called Milwaukee home and were hit hardest by things like economic downturns and unemployment. "What is the possibility," they asked, "of either government or private organizations creating new service jobs which use 'muscles more than brains?' Clearly, a pool of unskilled unemployed could be used to fill jobs in the private market which make it easier for people, with money, to get their chores done, e.g., snow shoveling, gardening." As befitting an organization dedicated to educational outreach, the committee suggested programming that spoke to these possibilities: teaching urban residents how to perform such tasks well.[38]

By 1974, 350 ten- by twenty-foot garden plots, sponsored by the Extension program, were placed in central city neighborhoods in Milwaukee. There was initially no charge for the plots, and they were located primarily on vacant lots owned either by Milwaukee County or by the city itself. There were also approximately 700 thirty- by thirty-foot garden plots sponsored by the City of Milwaukee (and overseen by the Department of Forestry) and neighborhood-improvement groups in operation. One prominent example of such projects was the City Garden Plot Program, which featured over 150 plots on city-owned land between Twenty-Fifth and Twenty-Seventh Streets and Lloyd and Brown Streets.[39] But, starting in 1976, users of the Extension gardens were charged $5.00 for a plot. Not surprisingly, this led to a decline in participation. In addition, Michael J. Brady, the chief field representative for the Democratic US representative Henry S. Reuss (who had become a staunch supporter of community gardening in Milwaukee), found: "One of the big problems that people had … is that they had gardened in the south but had never gardened in the north and needed much education as to what they could grow and when they had to have it planted in order to harvest good crops." At the end of the decade, a new organization, Shoots 'n Roots, would attempt to address both these issues.[40]

Shoots 'n Roots and Urban Gardening in the Inner Core

In March 1978, Earnestine Black, a resident of the city's North Side, wrote to Mayor Henry Maier, asking him to "give us pore [sic] people garden plot to raise food." In response, the city forester—and coordinator for the City of Milwaukee–controlled garden plots—Robert W. Skiera informed her that he had forwarded her letter to the new Shoot 'n Roots Urban Garden Program, which was "a part of the University of Wisconsin—Extension, Milwaukee County." He also included a brochure explaining the nascent program.[41]

Unlike other local gardening programs, Shoots 'n Roots had no plot fee. Under the banner "How to Save Money by Growing Your Own Food," the brochure explained that the program was "an educational gardening and nutrition program … to encourage limited income families to grow and use their own produce." It also noted that Shoots 'n Roots staff would educate participants on such topics as growing techniques, soil cultivation, and storing and preserving food. And the service was free: "You furnish the seeds and labor; we'll help you with the rest." What if a potential program participant did not have suitable space on which to garden? "No problem!" the brochure continued. "With the help of the Milwaukee Department of City Development, we'll help you locate a suitable garden plot."[42]

At the time of Black's letter to Mayor Maier, the Shoots 'n Roots Urban Garden Program was still getting off the ground. On December 6, 1977, the USDA announced that Milwaukee would receive $150,000 to fund the Extended Urban Garden Program for 1978. The funding, according to the USDA, was "aimed at helping low-income families grow their own vegetables," particularly on empty lots in inner-city communities. The pilot year for this program was 1977, when Congress allocated $1.5 million of USDA funds to urban gardening projects in New York, Chicago, and Los Angeles. The program was quickly deemed a success, and its budget was doubled to $3 million for 1978. For this year, Milwaukee, along with Atlanta, Baltimore, Boston, Cleveland, Jacksonville, Memphis, Newark, New Jersey, New Orleans, St. Louis, Philadelphia, Detroit, and Houston, received funding. The program was to be administered by the USDA Extension Service through the Cooperative Extension Services at the land-grant universities in the states where the cities were located.[43]

By mid-April 1978, Shoots 'n Roots projects had been initiated throughout Milwaukee, under the leadership of Urban Garden/University of Wisconsin—Extension

Agent Steven D. Brachman. By that time, Brachman had hired thirteen garden aides and three garden coordinators. In an April 1978 introductory letter to Mayor Maier, he wrote of the program's plan "to utilize vacant city land" in an effort to "encourage urban gardening and improve the nutritional level of low-income people." He also informed the mayor that he was working with Robert Skiera and the Bureau of Forestry to take over the administration of the city plot on Twenty-Seventh and Brown. After overseeing earlier city-led urban gardening programs, Skiera and his Bureau of Forestry associates were more than happy to cede control of such projects to Brachman; they did not have the labor force or the funding to commit to such additional work.[44]

Brachman also sent the mayor a "Plan of Work" for the fiscal year 1978. He laid out the situation in the eleven poor neighborhoods where Shoots 'n Roots was meant to do the majority of its work. Within such communities, rampant unemployment—more specifically, Black male unemployment—was perhaps the most pressing concern. "The 19.8% unemployment rate (November 1977) among blacks over 16 years old in Milwaukee," relayed Brachman, "is the highest rate among the 30 largest metropolitan areas according to the U.S. Labor Department's Bureau of Labor Statistics." For Brachman, Shoots 'n Roots could emerge as a way to get such individuals involved with productive labor once again. "It is apparent," he continued, "that one of the target audience groups for the Milwaukee Urban Garden Program should be the families of unemployed black workers." In addition to providing a form of work for such individuals, the program would also provide them a product that they could use, all the while taking some strain off their domestic budget. As Brachman concluded: "It would be imperative to concentrate efforts for people in these neighborhoods to learn to grow some of their own food by their own efforts."[45]

Such a process would also help instill healthy eating habits in a population that often suffered from malnutrition. As Brachman noted, Shoots 'n Roots would initiate the processes of "raising [program participants'] nutri-

tional level and realizing lifelong benefits from knowing how to raise and use nutritious foods." Participants could then share this information through educational outreach efforts aimed at other community members. Brachman even saw an entrepreneurial component possible within Shoots 'n Roots programming. He hoped that program participants would attempt to sell some of what they grew, and he also called for "the organization of a youth cooperative garden produce market whereby youth could learn methods of gardening and marketing."[46] While expectations for the program were not high, in its first year of operation Shoots 'n Roots reached over thirty-three hundred participants who helped plant over thirteen hundred gardens at twenty-nine community garden sites—a total of 200,000 square feet under cultivation.[47]

Such successful efforts to remake the city's landscape led Brachman to draw attention to another way Shoots 'n Roots was helping the city: its approach was a lower-cost way to deal with the problem of property dereliction in Milwaukee. Gardens, in other words, could do more than provide healthy food for city residents. They could also spur community redevelopment and help Milwaukee deal with a glut of vacant lots that were proving costly to the city. In January 1979, the *Milwaukee Journal* reported that, in 1978, the city spent more than $700,000 to maintain more than two thousand city-owned vacant lots. This broke down to a price tag of $346 per lot.[48]

Brachman directly referenced such figures in his appeal for city development funding in the winter of 1979 when he testified: "A primary factor I feel you should consider in your decision is the economic rationale for community gardening. Last year the city of Milwaukee spent over ¾ of a million dollars for the maintenance of 2,215 vacant lots." He informed the committee that, for newly acquired vacant lots, it cost approximately $1,500, or $0.55 cents per square foot, to prepare such spaces, which, "incidentally, were the lots most used as Shoots 'n Roots community gardens." In the Shoots 'n Roots proposal, he asked for approximately $33,000 of the $48,000 total request to fence and improve the soil of thirty vacant lots. This averaged out to approximately $1,100 per lot, which represented "a savings of approximately $400.00 over the maintenance cost of a newly acquired lot." Brachman also asked the committee to consider that the estimated yield of produce on an improved vacant lot exceeds $1,000 and that, once a community garden is established, "it requires minimum maintenance, while continuing to produce food for gardeners." For a city facing a shortage of funding, such possibilities were quite seductive.[49]

By the end of 1979, Shoots 'n Roots was recognized nationwide as a leading urban gardening program. In December 1979, the *Milwaukee Journal* reported that the Shoots 'n Roots program "aided 1,700 inner city gardeners, both in their backyards and on scattered vacant lots this year, up from 900 growers in 1978." That same year, the county (overseen by Richard C. Schneider Jr., the Milwaukee County horticulture agent) and the city (in the person of Milwaukee city forester Robert W. Skiera) made 2,100 and 950 garden spaces, respectively, available at seven locations. The value of the crops produced by Shoots 'n Roots gardens also increased. The gardeners involved in the program grew produce at a retail value of approximately $156,000, up from $92,000 in 1978.[50]

In order to publicize such efforts, Shoots 'n Roots staff began to organize tours of garden sites. For example, on July 25, 1979, a tour highlighted ten community gardens sponsored by the program.[51] Sites visited included a senior citizens garden where tour participants met with twenty-six gardeners, "all of them black." Other gardens were multigenerational. At the former Park West Freeway site, sixty-five-year-old Peter Docter gardened with the young mother Mary Mantyh, who set up a playpen for her children, Megan (six) and Casey (three). On the South Side, garden plots at the United Community Center and the Guadalupe Center reflected changing demographics—the former featured cilantro, while the latter had peppers raised from seeds brought from Puerto Rico.[52] The selection of these gardens to be a part of the tour was undoubtedly intentional, highlighting the fact that gardens reached those often overlooked by more

standard forms of urban redevelopment: minorities and the Elderly. Within these multigenerational spaces, the city's Black and Latinx populations were literally reshaping their neighborhoods.

Some took it further. During the summer of 1981, a group of young Milwaukee residents, aged seven to sixteen years, set up a farmers' market in downtown Carl Zeidler Park. The project was designed and supported by the Downtown Association, under the guidance of the University of Wisconsin—Extension. The participants were members of 4-H chapters, and many came from gardens sponsored by Shoots 'n Roots. At the same time, some who participated in the youth farmers' market were inpatients at the Child and Adolescent Treatment Center, part of Milwaukee County's Mental Health Center. For such young people, urban gardening was more than financially lucrative: it could also be both educational and therapeutic.[53]

As early as 1982, Shoots 'n Roots realized the potential for community gardening to bring about the creation of community—often across racial and ethnic lines. Gardens, as Brachman noted, could produce a much-needed product, but they could also help in "developing stronger neighborhood cohesiveness, providing neighborhood meeting places, and improving urban youths' awareness of their environment." Urban gardening, in other words, helped "develop neighborhoods."[54]

Notwithstanding these potential alternative positive outcomes, the focus remained on the financial aspects of community gardening as more and more observers came to take note of the continued growth of such projects. Across the city, the *Milwaukee Journal* reported, "gardening grows as economy falters." The *Journal* found that, of the 506,800 households in the Milwaukee metropolitan area, 268,000, or 53 percent, had a vegetable garden in 1981. In Milwaukee County, there were an estimated five thousand community gardening plots. This figure included twenty-six hundred rental plots at four locations on publicly owned county and city land rented and supervised by the county University of Wisconsin-Extension office, several garden plots on vacant inner-city lots under the supervision of Shoots 'n Roots, and other gardening projects run by church groups, by manufacturing plants, and by condominium homeowner associations.

The success of such individual gardens—along with the overall success of Shoots 'n Roots itself—led Mayor Henry Maier to declare May 5, 1983, as "Shoots 'n Roots Urban Gardening Day" for the city of Milwaukee. "What we have here, in essence," Maier concluded, "are small, family-oriented produce farms within the confines of an inner-city neighborhood."[55] By the mid-1980s, there were over twenty Shoots 'n Roots gardens spread across the city. According to one contemporary account, community garden production in Milwaukee was valued at $647,000 in 1985.[56]

Yet it was at this very moment that much of the funding for urban gardening programs in Milwaukee began to dry up. The economic recession of the early 1990s fundamentally altered the way institutions like the University of Wisconsin—Extension and programs like Shoots 'n Roots operated. As the *Milwaukee Journal* reported in May 1992: "In a recession economy, green things are easy to cut." Such things "are in danger in an era when more police officers and bigger jails are an easier sell." By that time, the city had already cut its number of full-time forestry employees, or those city employees who worked with urban gardening programs.[57] In the spring of 1994, Shoots 'n Roots announced that none of the twelve hundred gardeners in the program—at twenty-three sites throughout Milwaukee County—could continue to use watering hoses; they would have to haul their water by bucket or bottle to their garden beds. Of course, such a change in policy was effectively a death sentence.[58]

Growing Place: The Rise of Growing Power

As Richard Florida and his acolytes were using the early twenty-first century to highlight the relationship between the creative class and economic growth, a group on Mil-

waukee's North Side was drawing attention to the ways a vastly different demographic could embrace production as a viable means of urban redevelopment. In close proximity to a string of fast-food restaurants and nondescript strip malls—and a short walk from Westlawn Gardens, Milwaukee's largest public-housing project—Growing Power's two-acre tract of land at 5500 West Silver Spring Drive emerged as the most famous urban farm in America. The farm's ramshackle main building looked like an old barn; the *New York Times Magazine* described the entire compound as "an agricultural Mumbai, a supercity of upward-thrusting tendrils and duct-taped infrastructure." As Growing Power founder Will Allen has noted, aesthetics have never been his primary concern—production is. And there is no doubt that the business of Growing Power was food. The property remains packed with growing facilities, with Growing Power's main operation consisting, at its peak of production in 2015, of fourteen traditional greenhouses growing over twenty-five thousand pots of salad mix, arugula, beet greens, mustards, seedlings, sunflowers, radish sprouts, and a variety of herbs. The greenhouses were also the home for Growing Power's six aquaponic systems producing perch, tilapia, and an assortment of produce, not to mention more than fifty bins of red wriggler worms. Additional structures include two aquaponic hoop houses, seven hoop houses devoted to produce, a worm depository hoop house, an apiary with fourteen beehives, three poultry hoop houses, outdoor pens for livestock (including goats and turkeys), an elaborate composting system, an anaerobic digester, a rainwater catchment system, and a retail store. To walk through the property was to be awed by the unique sense of place the farm produced, to be overwhelmed by a series of sights and smells—yet still to feel the sense of calm often associated with more rural surroundings. One almost forgets that all this took place in the vicinity of a busy urban thoroughfare.[59]

In many ways, the rise of Growing Power was built on the histories outlined above. In the early 1990s, for exam-

A Growing Power employee harvests sprouts grown from the organization's aquaponic system.

ple, as Allen was beginning to grow food commercially, and before he founded Growing Power, he would sell his produce at the Fondy Market; he even worked to organize the vendors there into a cooperative. He also consulted with University of Wisconsin—Extension faculty members with years of experience in urban growing, including Dennis Lukaszekski, an Urban Agriculture Program coordinator and Community Gardens director who began working for the Extension in 1989. Yet Growing Power also served to address the evolution of these earlier histories. As the spaces sponsored and maintained by such programs as Shoots 'n Roots began to disappear

in the early 1990s, Allen came to see Growing Power—founded in 1993—as "filling the void." He notes: "If the Extension service had the funding and had the expertise of people, and I say this all the time to people, Growing Power wouldn't exist." Here, the government decision to decrease funding to the Extension brought about the need for an entrepreneur to step up and do the work that the public sector could no longer adequately perform.[60]

Such a misallocation of government power—as well as the subsequent response by the concerned citizen Will Allen—would make sense to Jane Jacobs and her acolytes. However, Growing Power emerged as a crucial actor in an

Growing Power's Will Allen
talking with visiting students
and other guests.

urban redevelopment narrative that challenged many of the tenets of contemporary placemaking, a narrative that offers another model of how to build place entirely. At the same time, Allen himself offers evidence of the presence of an intellectual and historical lineage that looks very little like the histories and ideas of those, like Jacobs, Richard Florida, or Ann Markusen, that also inform creative placemaking in the twenty-first century. His experience as the son of sharecroppers, along with his understanding of the lack of access to healthy food in urban communities of color, has led him to focus his work on the intersection of race, civil rights, and farming. And there is also an emphasis on this idea of production: Allen is intent on making his farm produce as much as possible. He is not interested in catering to the creative class or arguing the virtues of the postindustrial economy. He wants to make things.

A former professional basketball player and corporate manager, Allen quickly picked up the commitment to educational programming initiated by the Extension. He became a pioneer in both developing and teaching aquaponics (the cultivation of plants and fish together in a constructed, recirculating ecosystem) and vermicomposting (composting that utilizes various species of worms to produce fertile, nutrient-laden soil). Yet he also practiced traditional farming methods on a site that has been zoned for agriculture in the City of Milwaukee for over a century. The site—at Fifty-Fifth and Silver Spring—puts all these growing techniques predominantly on display. By 2011, the Growing Power farm was hosting over twenty thousand visitors and over four thousand volunteers. Through its on-site CUA training program and a variety of off-site workshops offered through the Growing Power National Training and Community Food Center Programs, Growing Power has trained hundreds of farmers who have gone on to set up their own sites in such US cities as Chicago, New Orleans, Detroit, and North Charleston, South Carolina, and countries as Haiti, Macedonia, Kenya, South Africa, Ukraine, and Zimbabwe.[60]

For Allen, the farm intimately connects the power of place with history (his own history and a broader African American past), his commitment to civil rights, and his desire to make something of value. Speaking on his desire to transform an underutilized space into a site of production, he comments: "I think, for me, it's kind of a legacy of my family. I can trace it back 400 years that we've been farming continuously. Through to my generation and now my daughter's generation." Addressing his own past— his parents had fled rural South Carolina to farm on the outskirts of Washington, DC—he recalls: "My family has always been a destination for feeding people. From the farm we were able to grow a lot of food—way more than we were able to eat and sell." More importantly, this food became a way to encourage sociability and grow community. "We used to share it with people, and my mother was a great cook," Allen continues. "At any time people would show up at our house in the evening, and we would feed them." Such recollections came to inform Allen's work at the Growing Power farm. He freely admits: "It really gives me a great pleasure to grow food for folks. From a plant, and really before that. Starting with the soil— understanding that it's all about the soil—and growing soil to be able to grow this food."[62]

At the Growing Power urban farm, then, the idea of placemaking can be taken literally. Through vermicomposting, Allen made the soil that now covers the property. The value of the place resides in the value of the actual land, and Allen has improved on this site through the addition of healthy soil, a rarity in a city where soil contamination due to industrial pollution makes urban farming problematic. On the one hand, such developments allow for Allen and those who interact with the farm to have a close relationship with it; they played an actual role in creating it. As they were starting this work in the 1990s, Allen recalls: "It was a very peaceful place for me—and for the people who worked there." But, through the constant process of remaking the site, he notes: "I still have a strong spiritual connection when I go in the building." At the same time, this brand of active placemaking allows Allen to see his work as speaking to broader concerns. "Food," as grown at sites like Growing Power, "is about social justice, environmental justice. That's what it's all about, done in a concrete way."[63]

Not surprisingly, then, Allen also sees the farm as speaking directly to civil rights. Around the farm, murals featuring depictions of the Great Migration and civil rights slogans such as "We Shall Overcome" and "I Am a Man" highlight the ways Allen connects his efforts to broader histories associated with civil rights in the United States. First and foremost, these murals depicting earlier freedom struggles allow Allen to connect his work to these earlier moments visually. He then uses the connection as a teachable moment, particularly for younger guests of the farm. "One of the best ways for them to understand the history," he explains, "is in a very visual way. So that's what that mural is kind of all about. Young people will be there and say, 'What's that?' They'll ask that question, and then you can explain to get them excited about that history."[64]

More specifically, Allen hopes that murals depicting civil rights movements displayed at a working farm will allow all African American guests to reconsider their relationship with the actual land and the profession of farmer. Once one reconsiders such histories, one can reconsider relationships that became troubled by the association of land and farming with slavery and sharecropping. "When you go back even further than that, to Africa," Allen continues, "where agriculture and growing food and taking care of your needs was part of the life—we come from a legacy, especially African American folks, of growing food. Long before any Europeans were really doing that." Ultimately, his work at Growing Power is "really about returning to your roots and getting young people to understand that history."[65]

But he wants to build on this history and find new ways to promote civil rights in the twenty-first century. It is with such ideas in mind that one should view a mural of a ticket taker beckoning observers to buy a train ticket. It is a nod

to the millions of African Americans who fled the horror of white supremacy in the rural South for hopes of economic opportunity in northern cities like Milwaukee. Allen sees Growing Power as doing this for the city; at its peak, Growing Power employed over 150 people, and that doesn't count the people who went through its training programs and started their own farms, companies, and other endeavors.

And all this was accomplished through a continued commitment to the land itself. This allowed the farm to become a true site of production. Each year, the farm produced thousands of pounds of produce—in 2013, for example, Growing Power grew and processed over 500,000 pounds of carrots, along with over 100,000 fish and millions of worms used to produce soil. This food was sold to grocery stores, restaurants, and schools. In 2014, Growing Power implemented the largest fresh food procurement in the history of the USDA-sponsored Farm to School Grant Program: 40,000 pounds of carrots sold to schools in the Milwaukee and Chicago regions. This produce was also distributed throughout the city through Growing Power's Farm-City Market Basket Program and sold at its on-site market as well as a variety of other farmers' markets throughout the city.

Such efforts put on display what Allen was growing while offering city residents increased access to affordable, healthy food. Yet they also put on display something more important: the ways urban agriculture can provide more than just food. "We want to show them that this could be a way for them to earn a living," Allen concludes, "because a food system has so many different jobs and job categories. We need finance people, we need renewable energy people, we need truck drivers. We could sit here and brainstorm a hundred different—almost every category of work that is connected to agriculture."[66]

To do this, Growing Power has worked with myriad universities, other nonprofit organizations, and government officials and agencies. Schools such as the University of Wisconsin—Milwaukee, Marquette University,

Aerial photographs of Milwaukee urban gardens and farms, including Walnut Way's orchards.

and the Milwaukee School of Engineering have provided Growing Power with faculty and service learners alike. Yet what is perhaps most noteworthy are the relationships that Allen has cultivated with civic leaders. Milwaukee alderman Don Richards, for example, helped Allen with zoning concerns in the early days, while the City of Milwaukee awarded Growing Power a $425,000 grant for the construction of 150 hoop house gardens on vacant land throughout the city in April 2011. Here, perhaps, one sees an example of the sort of government support for which Florida now calls.[67]

Beyond Growing Power—and beyond Milwaukee

The influence of Will Allen in Milwaukee—and, indeed, across the United States—cannot be overstated. Within Milwaukee, the strategies embraced by Allen have led others to employ urban growing as a novel approach to creative placemaking, pushing the work Allen pioneered in new and often unexpected directions. Perhaps the best example can be seen in Walnut Way Conservation Corporation, founded in 2000 by Larry and Sharon Adams to serve the North Side neighborhood of Lindsay Heights. This organization, with the stated goal "to reclaim and

redevelop the economic health and vitality of the community," started with a placemaking endeavor: transform a formerly vacant home into a vibrant community center. This house, located in the Lindsay Heights neighborhood, became Walnut Way's headquarters. Soon after, with the assistance of Will Allen, the group began growing—and, ultimately, selling—food in nearby lots. Commenting on this inspiration, Walnut Way cofounder Sharon Adams has noted: "Will both inspired and helped to lift the growing and marketing of gardens and orchards in Walnut Way." Perhaps more importantly: "Will encouraged and enabled us to become leaders."[68]

Early publications from the group stress this connection between Walnut Way and Will Allen. In the group's first-ever newsletter, from March 2001, in a piece titled "Community Gardens Help Raise Food and Spirits," it is noted: "Will Allen is pioneering the use of urban gardens for Milwaukee residents positively reusing vacant inner city land." Walnut Way should follow this example and "make a very positive use of some of the vacant lots in the neighborhood."[69] The next month, residents from the Walnut Way area attended an information session and planning meeting at Growing Power. At the farm, participants learned about aquaculture, growing produce, greenhouse construction, and vermicomposting.[70] A newsletter from July 2002 noted that children taking part in Walnut Way garden projects on Seventeenth Street would be "learning more about urban agriculture at the greenhouses of Growing Power … under the direction of Will Allen."[71]

The summer of 2001 saw the first full-fledged effort to get involved in gardening. By the summer of 2002, Walnut Way was producing and selling its vegetables, including fourteen hundred pounds of collard greens. In 2002, formal programs were started to get young people age eleven to fourteen into a twelve-week youth leadership and garden curriculum. This program used garden planning, harvesting, and marketing classes to teach natural science, math, writing, and time management.[72] Through such programs, participants planted one thousand tulip bulbs in October 2002. They harvested them in May 2003 and sold them for $10.00 a bunch. Proceeds went to support Walnut Way Market Gardens, through which "neighbors are transforming vacant city lots into beautiful market gardens."[73]

By the summer of 2004, Walnut Way had opened an outdoor market at Fourteenth Street and North Avenue, just blocks from its headquarters. This market featured produce grown in Walnut Way's rapidly expanding network of gardens, including tomatoes, peppers, and corn but also okra and "the largest collard greens this side of the Mason-Dixon line."[74] By the summer of 2006, Walnut Way was also growing and selling another crop—peaches, usually associated with the South—in a nearby vacant lot on the south side of the Walnut Way headquarters.[75] In 2007, the organization initiated the Gardens to Market Program. This initiative allowed the young people growing peaches and other crops to participate in an intensive urban agriculture internship program. Participants learned about growing methods and marketing techniques through hands-on programming and then were able to sell their produce at the Fondy Market. By 2017, Walnut Way was producing several hundred pounds of fruit annually, including peaches, cherries, apples, and pears, all being grown on formerly vacant city lots.[76]

As its growing peaches, okra, and collard greens illustrates, Walnut Way understands the role of history in its efforts. In addition to helping Lindsay Heights residents reconnect with their southern roots, the organization also wants to help participants reconcile their often-turbulent histories in the northern city of Milwaukee. And, as the 2010 organizational biography "The Walnut Way Story Project: Caring Neighbors Make Good Communities" notes, this process is place based. Walnut Way seeks to tell "the story of a neighborhood in Milwaukee's central city—what it used to be, what it became, and where it's going." Industrial growth throughout the twentieth century brought prosperity to the Black community in Lindsay Heights as working-class manufacturing jobs allowed residents to purchase homes within the community. Yet, between 1967 and 2001, Milwaukee lost 69 percent of its manufacturing jobs; a total of 82,178 positions vanished. As the Milwaukee economy moved from industrial to postindustrial, jobs quickly moved to the suburbs. To facilitate such movement, a highway system was "built right down the middle of Walnut Way," as the construction of Interstates 94 and 43 throughout the 1960s "tore apart the business district, uprooted people from their homes, and divided the community physically and spiritually, leaving a hundred vacant lots in its wake."[77]

This is the landscape and the history behind the creation of that landscape that, through its commitment to urban agriculture, Walnut Way is attempting to heal. Here, placemaking can be seen as regenerative, bringing new life back to once-abandoned spaces. Yet Walnut Way is now trying to regrow the economy as well—through adapted and new spaces. In 2015, the organization initiated phase 1 of its Innovation and Wellness Commons project, which centered on the adaptive reuse of a structure that was originally built in 1906 and has served as a bowling alley and a tavern throughout the majority of its existence. Walnut Way transformed this space into a multiuse structure, with a pop-up natural grocery food store, a juice bar—the Juice Kitchen, which makes some of its juices with produce grown by Walnut Way farmers—

Walnut Way volunteers and employees work on one of the organization's block-cleanup days.

The Sherman Phoenix's main hall.

and space for nonprofits operating in the community (including Fondy Food Center, the nonprofit that operates the Fondy Market). Food and urban growing are part of the story here, but this placemaking effort has expanded understanding of neighborhood wellness to include so much more.[78]

And the success of this $2.9 million redevelopment project has led to Walnut Way initiating phase 2 of the Innovation and Wellness Commons project. This phase will involve the construction of a new, 7,853-square-foot building, one that will continue the focus on health and wellness while also pulling in partners committed to job training, entrepreneurial opportunity, and educational programming. Tenant-partners included the African American Chamber of Commerce of Wisconsin, Milwaukee Community Business Collaborative/Jobs Work, North Avenue–Fond du Lac Marketplace Business Improvement District 32, and the Milwaukee School of Engineering, who will provide STEM (science, technology, engineering, and math) workshops for young people in the Lindsay Heights community. The phase 2 building will also feature a locally owned café and have space for local health, yoga, and massage providers. A 2,000-square-foot rooftop garden will serve as a convening space for community groups.[79]

In November 2018, Sharon Adams's daughter, JoAnne Johnson-Sabir, whose Juice Kitchen was an anchor tenant in the initial Walnut Way Innovation and Wellness Commons building, cofounded the Sherman Phoenix. The small-business incubator is housed in a former BMO Harris Bank that was damaged during the three-day uprising that followed a police shooting in the Sherman Park neighborhood in 2016. By early 2019, the site housed more than thirty Black-owned businesses, including 2 Kings Barber Shop, Next Level Vegan, and Amri Counseling, with the latter offering mental health and substance abuse counseling for children, adolescents, and adults. The building has been transformed from a visual reminder of violence into embodied community activism, and a young entre-

preneur with roots in urban agriculture and food now oversees a space employing dozens of people.

Others in Milwaukee inspired by Will Allen and Growing Power have a more complicated legacy. Cofounded in 2009 by James Godsil and Josh Fraundorf, Sweet Water Organics attempted to use aquaponics to turn a vacant industrial building on the city's South Side into a viable, commercial indoor urban farm. After having served on the board of directors of Growing Power, Godsil became enamored with Allen's approach to urban farming. "Will Allen," he notes, "was the direct inspiration for my work at Sweet Water.... [I] was committed to institutionalizing

James Godsil of Sweet Water Organics at the organization's former Milwaukee headquarters.

the charisma of big Will Allen." By 2010, the farm was selling about 150 pounds of produce a week and had sold over three thousand perch, all grown in vertical systems rooted in four eleven-thousand-gallon raceways dug into the concrete floor of the factory.[80]

Very quickly, however, Godsil realized that Sweet Water was more than a farm; it was a place that could be used for educational, social, and artistic programming. It was, in other words, a community center for the twenty-first century. "We viewed Sweet Water," he continues, "as an emergent, hybrid enterprise experiment, a social business and innovation center, advancing the commercialization, democratization, and globalization of aquaponics, an ecosystem method of food production." Yet, beyond being a center of "protein production," the farm became "a science lab, a school, an ecotourist destination, an artist and tinkerer's workshop, a community and new enterprise center." By 2009, it had created a nonprofit organization, the Sweet Water Foundation (SWF), to take the leadership on such processes. By 2011, it regularly hosted art shows, dance performances, and music concerts while continuing to offer programming for students from kindergarten through graduate school. Yes, Sweet Water hoped to produce a new generation of urban farmers, but Godsil was more excited about the space's ability to grow "beloved communities." Such language was intentional: Godsil had marched with—and provided security for—Dr. Martin Luther King Jr. during open-housing marches in Chicago in the summer of 1966.[81]

In May 2011, the city of Milwaukee approved a $250,000 forgivable loan to Sweet Water Organics. City officials, particularly the alderman for Sweet Water's Bay View neighborhood, were wowed by the facility's rapid growth. Yet, by late 2012, it was clear that the farm would not be able to meet the terms of the loan, tied as it was to an ambitious schedule of job creation. When it could not raise additional funds to expand production and pay back the loan, Sweet Water Organics closed in 2013. By 2017, the site was home to a new apartment complex, one that did little to preserve the history of either the farm or the industrial space that was there for generations.[82]

But the story of SWF has continued in Chicago. There, the foundation's executive director—and 2019 MacArthur grant winner—Emmanuel Pratt, who spent significant time in Milwaukee in 2008 to study the work of Will Allen as part of his dissertation for Columbia University's Graduate School of Architecture, Planning, and Preservation, has worked closely with city officials and community members alike to formalize this alternative approach. He has even given it a name: *regenerative development* or *regenerative placemaking*. For SWF, regenerative placemaking "is a creative and regenerative social justice method, that creates safe and inspiring spaces and curates healthy, intergenerational communities that transform the ecology of once-blighted neighborhoods." For Pratt, the integration of agriculture and art is central to the success of such a model.[83]

One sees what such an approach would look like in SWF's Perry Ave Commons, a multifaceted project launched on Chicago's South Side in 2014. The Commons features a three-acre farm that draws attention to the need to heal both land and history. "Food," the organization notes, "reconnects people to their origin—to the Earth and to each other. When grown in so-called 'blighted' communities, food is not only radical for these reasons, but also represents an extreme shift in habit, knowledge, policy, power, and place." Next to the farm is a community wellness hub—called the Think-Do House— housed in a foreclosed home. In 2017, SWF began constructing the Thought Barn, a barn that will provide space for the performing and visual arts while also providing a gathering space for community groups and members alike. And all this takes place on land that was given to SWF by the city of Chicago in a deal that was championed by Mayor Rahm Emanuel. Despite this relationship with city officials, Perry Ave Commons hopes to challenge "the standard reductive formula of gentrification, which seeks to gain short-term return on investment without regard to the longer-term sociological and ecological implications to the urban ecology, including people." Its approach is gaining admirers: in 2016, the group received a $300,000 grant from ArtPlace America to fund the Thought Barn. Such an award may suggest that even larger entities like ArtPlace America are seeing the placemaking potential of urban agriculture.[84]

Pratt's experiences in Chicago demonstrate how this story is taking root outside Milwaukee. In Detroit, D-Town Farm—founded in 2008 in the city's Rouge Park neighborhood by Malik Yakini—has its own close connection to Will Allen and Growing Power. Yakini, who also serves as chairman of the Detroit Black Community Food Security Network (a group that is attempting to encourage urban agricultural efforts throughout Detroit), is a graduate of Growing Power's CUA training program. Not surprisingly, D-Town's seven-acre farm shares many of the same characteristics of Growing Power. D-Town farm grows more than thirty different fruits, vegetables, and herbs that are then sold at farmers' markets and to wholesale customers. It also features four hoop houses for extended-season growing as well as large-scale composting to create their own soil.[85]

Yet Yakini and D-Town have taken something else from Growing Power. "Our affiliation with Growing Power," Yakini notes, "is particularly rewarding because we share the same commitment to access to good food and justice for all within the food system." As with Allen and Growing Power in Milwaukee, this means understanding the history behind such a food system. Detroit, like Milwaukee, was an early innovator in urban growing. In response to the economic downturn of the early 1890s—and its impact on recent Polish immigrants to the city—Detroit mayor Hazen Pingree began what the press called "the Detroit Experiment," "Potato Patch Farms," or "Pingree's potato patches." By 1895, over fifteen hundred families were growing food on vacant lots, producing a crop with an estimated value of $44,056. In a similar push during the Great Depression, the Detroit Thrift Gardens program, initiated in 1931, constructed 2,765 gardens on parcels throughout the city.[86]

Yakini sees his efforts as a part of this story. But, like Allen, he also sees his work as a part of the civil rights narrative: "If you're not aware of the last 50 years of struggle for black empowerment in the city of Detroit . . . then you're likely going to move in a way which is not going to be supportive of community goals and aspirations, and perhaps even undermine those." Because of this, he sees D-Town as "definitely rooted in the social justice, racial justice, economic justice camp . . . making sure that the profits from that food is circulated within our community as opposed to being extracted."[87]

By late 2017, Yakini was working, through the creation of the Detroit Food Commons, to bring these ideas to a new level, one akin to the work that Walnut Way is doing in Milwaukee. The Commons will be a new thirty-thousand-square-foot campus in the North End that will include the Detroit People's Food Co-Op (a member-based cooperatively owned grocery store), a café, a community meeting space, and an incubator kitchen. Yakini is also working with Develop Detroit, which will provide sixty-five units of affordable housing for the project. Groundbreaking for the grocery store is scheduled for 2021.[88]

There is little doubt that Yakini sees this project as producing things to meet specific community needs. Speaking about the Commons, he has noted: "It has the kind of mix that we were looking for where we can serve the need of folks who don't have abundant access to multiple options for obtaining good food." But, more broadly, he sees the project as capable of positing an alternative understanding of both urban redevelopment and place. "We want to be on Woodward Avenue," he concludes, "to make that statement in the face of development happening in downtown or so-called 'Midtown' that is largely the kind of typical development that we see in the United States led by wealthy white men, where communities really don't have much of a stake in the ownership."[89]

In New Orleans, another urban farm with a connection

Grow Dat Youth Farm Leadership Program participants work at the
farm in New Orleans.

to Growing Power—Grow Dat Youth Farm—is working to address similar concerns. While Grow Dat founder and former executive director Johanna Gilligan cites Urban Roots in Austin, Texas, and the Food Project in Boston as the primary inspirations for her project, she was a graduate of Growing Power's first CUA training program and remains in contact with Will Allen. Founded in 2010, Grow Dat's two-acre farm was by 2017 employing scores of young people through its leadership programs and producing more than twelve thousand pounds of produce. Grow Dat sells 70 percent of this produce at farmers' markets run by their young employees and distributes the remaining 30 percent to low-income New Orleans residents through its Shared Harvest program. It has a community-supported agriculture program that is made up of nearly one hundred members who pay for weekly produce boxes in the spring and fall. It also runs two spring markets and hosts farm dinners, site tours, and other community-oriented events. Commenting on this activity, Grow Dat "hive coordinator" Jabari Brown notes: "It's a work of art, man, the magic that happens."[90]

Brown is also quick to note the importance of Tulane University in the development of Grow Dat. "We have a partnership with Tulane," he says, "that's really gracious. I mean, they literally fostered us for the year.... We're clearly indebted to that." Indeed, Grow Dat was incubated by Tulane and Tulane City Center and continues to receive support from the institution. During a year-long fellowship from the institutions, Gilligan built on her experience with the New Orleans Food and Farm Network (NOFFN) and a long-held idea to "hire teenagers to grow food," paying them to increase access to quality food and develop leadership skills. Pairing the university's interest in community engagement and Gilligan's connections, the group piloted the program for its first year at NOFFN's Hollygrove Market and Farm, an urban agriculture experiment and neighborhood-based market for locally produced food. During that first year, Grow Dat drew from its university resources and negotiated a cooperative endeavor agreement with the city's storied City Park to find a permanent home for the Grow Dat Youth Farm "eco-campus."[91]

As Gilligan designed programming for the new site, she coordinated with the university's Corporations and Foundations Relations Office, design studios, and local high schools to strategize for the new facility. One of the unusual funding sources came from John and Anne Mullen. The Mullens were interested in using damaged shipping containers no longer suitable for transport. The new facility prominently features such containers in a mix of indoor and outdoor spaces, a fitting solution from John Mullen, the architect of the original Container Store. Since moving to City Park in 2012, Grow Dat has settled into the site and is maturing as an organization. Even as Hollygrove Market and Farm has faltered, closing amid declining sales and other financial issues in early 2018, Grow Dat is expanding and moving toward a more independent future.[92]

In Cleveland, Mansfield Frazier has been inspired, in part, by Allen to open Chateau Hough, a vineyard in the Hough neighborhood. Since the 1960s Hough has been a symbol of segregation and impoverishment, in part owing to its history as the site of the Hough rebellion, a precursor to the "long, hot summer" of 1967. By the early years of the twenty-first century, decades of divestment were paired with the foreclosure crisis, leading to a landscape of abandonment. According to one estimate, there were six thousand demolitions of abandoned buildings from 2009 through 2013. At the same time, there are close to four thousand acres of vacant land in Cleveland, with nearly half held by the city. The neighborhood remains one of the neighborhoods with the highest concentration of unoccupied and derelict properties in the country.[93]

Frazier built his own house in Hough in 2000 and invested another $200,000 in a six-home development in 2007—right before the housing bubble burst. Determined to continue his community-improvement efforts, he started the vineyard on reclaimed land in 2010 as

The Vineyard of Chateau
Hough in Cleveland.

Indianapolis; Milwaukee; Chicago; Washington, DC; Milwaukee; San Francisco; Phoenix; Denver; Las Vegas; Buffalo; Washington, DC; Braddock, Pennsylvania.

the foreclosure crisis started to ease. The farm was initially funded through a $15,000 grant from the city and close to $8,000 of Frazier's own money, along with advice offered by the Ohio State Extension Service. Frazier felt that a vineyard could add value to his property while kick-starting redevelopment within the broader Hough community. And it made a statement that a typical farm would not. "If I'd planted bell peppers," he told a reporter from the online publication *Good*, "you wouldn't be standing here."[94]

In addition to being a symbol of wealth, wine can be sold for more than the average garden product. The grapes, grown in 294 vines, set in fourteen rows with twenty-one plants per row, are of the Frontenac and Traminette varieties, both of which grow well in cold weather. While the finished products are not among the elite of the wine world, the vineyard's white wine did win a second-prize ribbon at the Cleveland Geauga County Fair in the summer of 2014. Chateau Hough received full certification to distribute its wines in 2017, by which time it was producing over one thousand bottles a year in six different styles. The wines can be bought in Cleveland-based Heinen's grocery stores across the city and in restaurants in such neighborhoods as University Circle.[95]

Along with the vineyard, Frazier built a biocellar, a below-grade growing environment housed in the rebuilt basement of an adjacent, razed abandoned home. The temperature within the biocellar remains constant at approximately fifty to fifty-five degrees Fahrenheit throughout the entire year, allowing for an extension of the growing season. At the same time, the structure reuses embodied heat from the house itself and the natural insulation offered by the earth surrounding the foundation. As with the vineyard, Frazier sees the main goal of the biocellar as wealth creation. He plans to grow a variety of foods in the structure, starting with mushrooms. "All of what we do is wealth creation," he notes, "and we have to grow what we can grow for the highest dollar amount. . . . Shiitake mushrooms are $12 a pound." Taking a cue from Grow-

ing Power, Frazier wants to employ aquaponic systems for fish production within the biocellar.[96]

In addition to wanting to revive spaces to make them more productive, Frazier sees Chateau Hough as doing something similar for people. He himself is an ex-convict—in 1992 he was sentenced to three years in federal prison for credit card fraud—and he sees the vineyard as offering recently incarcerated men a chance to work. Much of the work on the construction of both the vineyard and the biocellar, for example, was done by residents of a neighborhood halfway house and work crews made up of individuals performing court-imposed community service. "Our job," Frazier concludes, "is to solve poverty with paychecks."[97]

In North Minneapolis, Appetite for Change is following the formula for greening urban space that others have been adopting in other midwestern cities like Detroit and Milwaukee. Through its Fresh Corners Initiative, the organization works with a collection of ten farmers working on thirteen sites in North Minneapolis to get their produce into corner stores, area restaurants, and farmers' markets. Through its Community Cooks program, Appetite for Change has held over 150 workshops on growing and preparing healthy food, nutrition, and food justice. All this activity comes together in the group's Kindred Kitchen, a shared commercial kitchen that allows urban growers to process their produce and gives food trucks, caterers, and other local food providers the space to prepare their meals. The kitchen also offers classes to community members who are planning on entering the food industry.[98]

Yet Appetite for Change has become best known for its use of hip-hop in spreading its message. In November 2016, a group of Appetite for Change interns—in collaboration with Beats and Rhymes (a nonprofit founded in 2007 to provide school-age children the opportunity to record their own music)—released a rap song called "Grow Food" that quickly went viral. In addition to exhorting their listeners to "grow food," the talented rappers spit rhymes critiquing the physical layout of their neighborhoods while highlighting the lack of healthy food in many communities of color and the impact that such developments have on community health. "See in my hood there ain't really much to eat," raps one young man. "Popeye's on the corner, McDonald's right across the street / All this talk about guns and the drugs, pretty serious / But look at what they feeding y'all, that's what's really killin' us." Their response to this? "Get that fake food up out of my hood."[99]

But "Grow Food" is more than a critical wake-up call. It is also a positive call to action that stresses individual agency and entrepreneurial potential. "Grow and cook your own food, yes you could," the song continues. "Got the lil homies in the garden / Got the big homie sellin' collard greens." It is hip-hop culture speaking to urban agriculture in a way that reminds the listener of the power of art to transform both minds and places.[100]

The Place of Placemaking

The history of urban agriculture in Milwaukee clarifies that the connection between community development and place expressed in contemporary placemaking is and always has been embedded in the issues and organizational actors of the day. Moreover, this historical approach illustrates how current successes and failures of urban agriculture's intentionally place-based approach to community cohesion, food production, workforce development, and civil rights activism are rooted in this history, a history that looks vastly different from the histories often privileged by placemakers and placemaking organizations in the early twenty-first century. When a guerrilla gardener plants seeds on a derelict lot, she is telling not only the stories of the creative class or the Floridian city but also the broader story of urban engagement by its marginalized residents, often in coordination with small-scale, not-for-profit organizations as well as local, state, and federal governments, all of whom are often written out of the history of contemporary placemaking.

The resulting story is one of direct engagement with the economic and social contributions of siting agriculture in the city amid the dramatic demographic changes that occurred in Milwaukee during the second half of the twentieth century. The reorientation of urban growing followed the already-extant idea of providing healthy food and employment opportunities in underserved communities and combined it with racial and cultural empowerment. The result was new models for interorganizational cooperation in an explicitly racialized and politicized environment. Of course, this counterhistory did not develop in isolation; it existed side by side with the narratives of urban renewal and critiques of such strategies as put forward by the likes of Jane Jacobs, the New Urbanists, Richard Florida, and others concerned about the late twentieth-century city. What is worth stressing here, however, is how such urban agricultural endeavors came to challenge the belief that successful placemaking efforts must take the city as postindustrial for granted. There is a story of consistent urban production on display at such sites, one that draws attention to the relationships between place, labor, and product in ways that push the placemaking conversation away from the creative class and its allies.

The next two chapters of this book will be informed by the lessons gleaned through this thorough examination of urban agriculture in Milwaukee and other cities. As such, they will not devote the same space to the concept of urban agriculture as placemaking or to enumerating contemporary groups' historical time lines and contexts. Rather, with a continued emphasis on those often left out of the mainstream placemaking story, we will see how a variety of manufacturers, companies, nonprofit organizations, artists, and activists are rethinking this relationship between place and production in innovative ways. It is in such work that we see a further challenge—indeed, a useful challenge—to the ways creative placemaking has been historicized, articulated, and practiced in the early twenty-first century.

Producing Place

(*Previous spread*) Visitors tour the Brooklyn Grange, a major rooftop
farm in New York City.

As we chronicled in the last chapter, urban farms are transforming the urban landscape and offering city dwellers new understandings of place and placemaking. These relationships are rooted in long-standing practices, but there also is little doubt that a place like Chateau Hough challenges our understanding of what a city—and particularly a troubled neighborhood in a city hit hard by disinvestment—can look like. If nothing else, such efforts suggest that, in cities like Cleveland, Milwaukee, and Detroit, there is now an opportunity to reintegrate nature and farming into the planning of the city and that a variety of individuals and organizations are doing such work. Moreover, a hallmark of this opportunity is how government and private organizations can cooperate with activists to redistribute land ownership in central cities and, in the process, facilitate the production of wealth in communities where it has been eroded in recent decades.

But urban agriculture can introduce another element left out of the placemaking discourse while transforming the ways cities think about jobs, labor, and urban redevelopment in what many describe as the *postindustrial era*. Even in a city like New York, which is not only the epicenter of the work of Project for Public Space (PPS) but whose economic growth has insulated it from some of the worst effects of the financial crisis, urban farmers are contributing new approaches to making underutilized places economically productive again. For example, Annie Novak and Ben Flanner, who grew up in the Milwaukee metropolitan area and were inspired by the efforts of Will Allen and Growing Power, established the Eagle Street Rooftop Farm there in 2009. The farm is made up of 200,000 pounds of growing medium spread across a six-thousand-square foot green roof, allowing for the cultivation of over thirty different crops. And, while such efforts may seem vastly different from what the more activist-oriented Appetite for Change has done in Minneapolis, the example of Eagle Street suggests just how widespread this desire to rethink something like urban

food production has become. It also highlights the significant impact such a rethinking can have on the urban built environment. The green roof design and installation was done by the firm Goode Green and illustrates how the urban farms of the twenty-first century are embracing innovative, eco-friendly design concepts. As one might expect, this roof is a sight to be seen: a lush green space in the midst of a concrete city. Yet equally as noteworthy is the breathtaking view provided by the farm's rooftop location: both the East River and the Manhattan skyline.[1]

At the end of the 2009 growing season, Flanner left Eagle Street to cofound another urban farm: Brooklyn Grange. Joined by Gwen Schantz, Chase Emmons, and Anastasia Cole Plakias, he looked to farm in neighborhoods affected by deindustrialization: the first Brooklyn Grange farm opened on a rooftop in Queens in May 2010. Measuring one-acre—or forty-three thousand square feet—the Flagship Farm is made up of close to 1.20 million pounds of soil. A second farm, the Navy Yard Farm, opened on a rooftop in Brooklyn in 2012. This farm covers sixty-five thousand square feet, the equivalent of 1.5 acres. These two farms produce over 50,000 pounds of produce each year; they have sold over 500,000 pounds of vegetables to a variety of individuals, restaurants, and other businesses. Brooklyn Grange also operates an apiary featuring over thirty hives spread out on rooftops throughout New York City. The two farms employ fifteen full-time employees and over forty part-time and seasonal staff members. They run two weekly farmers' markets, manage vegetable and flower community-supported agriculture programs, and sell directly to a large number of retailers, restaurants, and catering firms. As Brooklyn Grange cofounder Anastasia Cole Plakias notes in her 2016 *The Farm on the Roof*: "We didn't simply want to farm; we intended to create a small farm business—a self-sustaining enterprise that, like any other business, would have to turn a profit to survive." But they also saw the project as something that could affect the urban built environment, "an industry that, if invested in, could help change the land-scape of cities."[2]

Such a quote suggests that Plakias and her colleagues are not your average capitalists. In fact, she makes it clear that their efforts were a response to the opportunity provided by the Great Recession to rethink things like economics and labor. Rather than seeing that particular moment of economic uncertainty as a negative, Brooklyn Grange saw it as "a time marked by potential": "It was a moment during which revolution seemed possible." Plakias is quick to note that such economic "disrupters" as Airbnb (founded in 2008), Uber (2009), and Kickstarter (2009) were all founded at approximately the same time as the first Brooklyn Grange farm. In fact, Brooklyn Grange was an early adapter of crowdfunding, using the Kickstarter platform to raise money for its Flagship Farm. "Who knows if Brooklyn Grange would ever have received the support that it did—from investors, from the larger community, even from us, its founders," concludes Plakias, "if there hadn't been such a spirit of transformation, a shift in public perception of what was and wasn't possible."[3]

Yet Plakias is also quick to note the physicality of Brooklyn Grange's act of economic disruption—and that it is more than just a phone- or web-based application that is designed to please shareholders. As the Great Recession gave way to Occupy Wall Street and then to concerns regarding climate change and racial upheaval, Plakias came to see her farm as part of a broader urban community, one that reconceptualized both farming and business/economics in new ways. In an epilogue titled "Toward a Healthier Business Ecosystem," she notes that the farm exists within an *ecosystem*, which she defines as "a group of interconnected elements, formed by the interaction of a community of organisms with their environment." The farm feeds its customers and works with the for-profit businesses and nonprofit organizations that it houses, which in turn enrich the business of the farm. The city provides economic incentives, and the farm then provides green infrastructure that is both beautiful and functional: "We all support one another; we all work better because of our interconnections."[4]

Practically, this has meant that funding has come from loans, fundraising events, private equity, and online crowd-funding campaigns, which have provided new ways for groups to build audiences and raise money from smaller donors. New York City, primarily through its Department of Environmental Protection (DEP), has also contributed funding. Through its Green Infrastructure Stormwater Management grant program, DEP awarded Brooklyn Grange over $590,000 in 2011. Of course, the DEP appreciated that the green roof created by the farm would manage over one million gallons of stormwater per year and reduce Combined Sewer Overflows (CSOs) to the East River. But city officials saw more to the project than cleaner rivers. While stressing that such environmental outcomes should be welcomed, Chris Tepper—Deputy Director of Development and Planning for the Brooklyn Navy Yard Development Corporation (BNYDC)—saw even greater value in the fact that the farm would "produce fresh produce for surrounding communities and meet BNYDC's core mission of creating jobs." Finally, Tepper was excited that a type of production, and green production at that, was coming to the city. "This partnership," Tepper concluded, "is another step towards redeveloping the Brooklyn Navy Yard as the greenest industrial park in the country."[5]

This notion of urban agriculture as a viable means of both job creation and economic development is spreading across the country. In Chicago, the public City Colleges system now offers an advanced certificate in sustainable urban horticulture, while Kennedy-King College—a member of the Chicago City Colleges consortium—regularly runs courses on aquaponics. One of Kennedy-King's partners in the development of its urban agriculture curriculum has been Growing Home, a Chicago-based nonprofit that, since 2002, has offered job-training programs geared toward the organic agriculture industry. Growing Home has made it a point to tailor its programs to those often left out of the labor pool, including the homeless, recover-

Growing Home farm manager Deandre Brooks flames the beds to kill chickweed seeds organically.

ing drug addicts and alcoholics, and the formerly incarcerated. At its Wood Street Urban Farm (which opened in 2009) and its Honore Street Urban Farm (2011), it provides such individuals twenty-five hours per week of paid on-the-job experience and job-readiness training. Other Chicago-based organizations have taken inspiration from such urban agriculture endeavors. Sweet Beginnings, an urban apiary in the North Lawndale neighborhood of Chicago, is setting up beehives on underutilized properties throughout the city and providing job training to similarly economically dislocated city residents. A subsidiary of the nonprofit North Lawndale Employment Network, it harvests these hives to produce goods both edible (an all-natural line of raw honey) and nonedible (honey-infused body-care products). Most importantly, it provides full-time transitional jobs for formerly incarcerated individuals and others facing real barriers to gainful employment. These workers care for the bees and hives and harvest honey, among other tasks. The recidivism rate for former Sweet Beginnings employees is below 4 percent, compared to the national average of 65 percent.[6]

This chapter uses these projects as a jumping-off point for a broader discussion of how such efforts are leading to a reimagining of the relationships between place, labor, and economy in the twenty-first century. The act of growing food highlights what is perhaps most striking about this contemporary moment and why we see it as so important for any discussion on place and labor in the twenty-first-century American city. Throughout the country, we see the return of production, of people actually making things. Actors as diverse as community gardeners, food innovators, do-it-yourself (DIY) tech entrepreneurs, independent business owners, and small-scale manufacturers are challenging the myth of the postindustrial city. Importantly, these efforts are led not only by creative-class migrants but also by long-term residents of communities often left out of mainstream creative-placemaking efforts. Perhaps not surprisingly, then, these individuals and organizations are often committed to selling their goods in new ways or providing services in innovative ways and/or novel locations. The economic collapse of 2007 provided the resources—including abandoned buildings, vacant lots, and a population base desperate for work—for all of this to happen. In turn, the growth of a new breed of service-providing industries is demonstrating that opening a store or providing a service for a neighborhood should be seen as both a political act and a means of building much-needed community through a novel approach to placemaking.

It must be noted that many of these efforts are not always welcomed by city governments, and, in fact, many of the endeavors chronicled in this chapter could not have gotten off the ground if they had adhered to all government regulations and guidelines. Yet, understanding that they may be able to take advantage of evolving definitions of work and economic development, some cities are more conducive to such projects than others. On one level, this is often due to a simple lack of resources. Detroit, for example, does not have the money to enforce zoning and property laws consistently. However, other cities like Milwaukee, Chicago, Cleveland, and Kansas City, Missouri, have governments that have become actively engaged with the creation and promotion of alternative

growth strategies. And, while it would be folly to suggest that these efforts have been fully realized, one cannot deny that change is coming to cities across the nation, change rooted in how we work and how we think about such work.

Ultimately, this chapter seeks to document what can best be described as *placemaking as alternative economic development*. At the heart of such a place-based approach to development is a commitment to making things, even in cities that are often described—and readily dismissed—as *postindustrial*. Such a return to production has also led to the return of certain types of jobs to American cities as well as the creation of new ones. Of course, the mainstream approach to creative placemaking offers similar rewards. But what we see in this chapter looks quite different from the creative class fueling urban regeneration in the schemes of individuals like Richard Florida and groups like the National Endowment for the Arts (NEA). As opposed to their arts-centric strategies, here we privilege efforts that are making physical things. But, perhaps more importantly, we are seeing a rethinking of what work actually entails. This involves not only a fundamental rethinking of the relationship between jobs and labor but also a rethinking of such concepts as commodity, service, exchange, and value as city residents strive for a more holistic understanding of economic life. These radical realignments allow, we suggest, for a more inclusive understanding of creative placemaking, one, once again, more attuned to the realities of both race and class. If nothing else, the sites we highlight in this chapter offer jobs and services for those often left out of any discussions of the creative class and creative placemaking while creating new avenues through which social capital can grow and flourish.

Our conclusion that creative placemaking can drive both economic and spatial redevelopment is not novel, but the realities of abandoned and underutilized places, particularly following the havoc wrought by the Great Recession, created the necessary space, literal and theoretical, to allow these trends to touch the lives of more and more city dwellers; reclamation of vacant or under-utilized space becomes an act of asset redistribution. And efforts to transform these sites into productive, inclusive places have had the effect of reinvigorating them while simultaneously shoring up, through the creation and maintenance of a number of new and previously existing social relationships, the urban fabric that surrounds them. As these projects enter communities hit hard by the reali-ties of state-sanctioned urban renewal efforts, industrial pollution, riots and uprisings, disinvestment, and popula-tion flight, they draw on a set of histories and intellectual influences and traditions that mainstream placemakers often overlook. As we began to see in chapter 4, some of these continue to draw from a rich history of civil rights activism, while others draw from alternative understand-ings of community development and economic activity. Some even bring matters of environmentalism, spiritu-ality, and religion into the picture. Yet, from deindustrial-ized communities in Detroit to rapidly changing neighbor-hoods in Brooklyn, these projects share a common goal of seeking to heal the damage wrought on these places by previous economic-redevelopment plans. Sometimes this act of healing is profoundly environmental; other times it seeks to heal broken social bonds. But all of it is part of a broader process of calling for—and working to create—new understandings of labor and economy.

And government officials and other policymakers are paying attention. As the example of Brooklyn Grange sug-gests, the potential to bring production back to a commu-nity hit hard by disinvestment, for example, has proved attractive: government and philanthropic funding is now being offered for such projects, and zoning ordinances are being rewritten. But what happens when many of these alternative, perhaps even oppositional approaches to ur-ban economic redevelopment gain the legitimacy offered by the state? In a number of postrecession American cit-ies, these underground approaches to placemaking are beginning to enter the mainstream.

Production Strikes Back: A Post-Postindustrial Economy?

This story of production in New York City is not solely a tale of urban agriculture. One sees similar initiatives on display while walking through the industrial sections of Brooklyn—where things are still actually made. Once ensconced in the factories and small trade shops of the Greenpoint neighborhood, it is easy to forget that both hipper-than-thou Williamsburg and center-of-the-global-economy Manhattan are so close. On a warm September day, the New York–based artist/filmmaker Sarah Nel-son Wright shows us around this part of Brooklyn, point-ing out the local companies—Acme Smoked Fish, Royal Engraving, Associated Fabrication, and Dobbin Mill—she documented in her 2009 site-specific video installation *Brooklyn Makes*, a project that examines the industrial past, present, and future of the borough. The film, which was projected on the sides of buildings in the Williams-burg Greenpoint Industrial Zone, is meant to "reveal dynamic and essential aspects of the city that often op-erate outside of the public eye": "*Brooklyn Makes* reveals functional industry coexisting in a residential neighbor-hood that is thought of as postindustrial."[7]

And things are being produced here—in ways that pro-vide jobs for the community at large. Acme Smoked Fish, which traces its history back to the 1905 emigration of Harry Brownstein from Russia to Brooklyn, engages in the production and packaging of smoked fish and employs over 150 workers, including many recent immigrants. Royal Engraving is a traditional engraving workshop that opened in 1978 and employs approximately 25 people, while Associated Fabrication is a full-service digital fab-rication and architectural millwork company that serves contractors, architects, furniture makers, and others. Yet not all Brooklyn's manufacturing firms are so industrial: Dobbin Mill, founded in 1989 by Robbin Ami Silverberg, is a hand-papermaking studio that also features an artist-book studio, a courtyard for outdoor work, and a paper-

The kind of scene typically promoted in North Brooklyn's industrial districts at the time of *Brooklyn Makes*.

maker's garden: the trees that make the paper for Silverberg's artwork and artist-book collaborations under the Dobbin Books imprint are grown on-site. There is little doubt that Silverberg's space is more than just a site where things are made; it is a place. And, while Brooklyn has obviously changed since Wright made her film, all the firms chronicled in her work remained open in 2019.[8]

All these firms predate the financial crisis, but Wright's decision to document their existence was tied directly to the worsening economic climate, along with the belief that the early twenty-first-century American city no longer made things. According to Wright, the economic downturn "seemed like the right moment to point out

these businesses, which do not enjoy much support from the city and are under constant threat of dislocation." At the same time, the Great Recession "caused questioning of global economic systems" and led to some wanting to move beyond simply "lamenting the loss of local manufacturing in urban centers." For Wright, then, an exploration of how Brooklyn's manufacturing past remains relevant to the present could provide a new way "of imagining the future" of the neighborhoods once written off as postindustrial.[9]

One gets a view of what this marriage of manufacturing past, present, and future might look like in Chicago. On first sight, The Plant—a self-described "collaborative community" of small food businesses—looks like a remnant of Chicago's old economic order. The imposing brick building in the city's Back of Yards neighborhood was once the home of Peer Foods' meat-processing and -packing plant. In the immediate post–Civil War era, this neighborhood became known for its multitude of stockyards and slaughterhouses as it used the city's extensive railroad network to become the chief supplier of meat for the United States. By 1868, the Union Stockyards, the city's attempt to centralize such operations, had grown to consist of approximately twenty-three hundred pens housed on one hundred acres. It was in this space that Buehler Brothers, as Peer Foods was initially called, opened its packing facility in 1925; the company would change its name in 1944 and continue to grow throughout the twentieth century.[10]

Yet advancements in technology, transportation, and distribution methods in the late twentieth century ultimately brought about the demise of Chicago's meatpacking district. The rise of interstate trucking, for example, allowed for the direct sale of livestock from breeders to packers, while the development of refrigerated trucks meant that facilities no longer had to be close to the railroads. This allowed companies to move their slaughterhouses to rural locations, where both land and labor were cheaper. Perhaps most importantly, these new locations, often in the American South, allowed companies to move away from the influence of Chicago's powerful labor unions. By the early twenty-first century, the Back of the Yards had lost the majority of its meatpacking plants. Peer Foods itself finally closed its doors for good in 2006.[11]

The building sat vacant until July 2010, when it was purchased for $525,000 by the Chicago-based entrepreneur John Edel. While the original plan was to sell off the materials inside the mammoth structure, Edel was intrigued by both the building's history and its seemingly indestructible nature: you could put anything, regardless of weight, on any of its four stories. Edel envisioned the 93,500-square-foot site as a hub for urban agriculture and food production and christened it The Plant. Aqua-

ponic systems were put in its basement, while outdoor spaces were transformed into lush gardens. Soon, food producers throughout the city began to set up shop within: by August 2013, it was home to Peerless Bread and Jam Company and the kombucha maker Kombuchade, among other tenants. A neighborhood once known for slaughterhouses and tanneries was now home to a new hub of production.[12]

Yet there was more behind Edel's motivation than profit. By starting The Plant in an area hit hard by deindustrialization, Edel hoped to create jobs for the economically distressed neighborhood. At the same time, the flight of businesses that went hand in hand with such disinvestment had transformed the Back of the Yards community into a food desert. The Plant could provide healthy food in an area that sorely needed it. And all this was to be done in a sustainable, environmentally friendly manner, with the goal of healing the legacy of industrial pollution and waste. The building is meant to provide a physical example of a circular economy, one in which energy needs and waste cycles are dealt with internally, with the goal of becoming a zero-energy, zero-waste facility. To that end, in September 2011, The Plant was awarded a total of $1.5 million in grant funds from the Illinois Department of Commerce and Economic Opportunity to support the construction of an anaerobic digester that will turn up to twelve thousand tons of food scraps and other waste produced in the building and the surrounding area into energy. Completing the project has taken a backseat to the rehabilitation of the building itself. As of 2019, the digester is complete, permits have been issued, and the site is prepped for installation in 2020. Once online, it is expected to help The Plant move off the traditional power grid completely.[13]

Such innovations have served as compelling recruitment tools as each year The Plant adds more tenants to the building. By 2017, the building was home to approximately eighty full-time employees working in a variety of food-producing endeavors. The list of tenants had grown to over fifteen, including companies like Tuanis Chocolate,

The Plant and its garden.

The demolition of a former Back of the Yards brewery.

the Great American Cheese Collection, Bike a Bee (which produces honey products), the ice cream–maker Sacred Serve, the Whiner Beer Company, and Rumi Spice. There is little doubt that such companies are evidence of the site's commitment to cultivating a hyperlocal food economy. Yet the example of Rumi Spice shows the potential global reach of the site. Rumi was founded in 2011 by the army veterans Emily Miller and Kimberly Jung, who, motivated by their tours of duty in Afghanistan, sought to provide an economic opportunity for Afghan farmers to grow something other than poppies. It now works with over thirty farms—which employ approximately four

hundred women—to bring the raw materials needed to create spices such as saffron from Afghanistan to the United States.[14]

What is perhaps most innovative about The Plant, however, is that it developed nearly in tandem with a nonprofit organization called Plant Chicago. Plant Chicago was created in 2011 and came to serve as the steward of the building, whether internal or external relations were involved. And, while Edel initially wanted it to focus on research connected to The Plant's vision of a circular economy, current Plant Chicago education and outreach manager Kassandra Hinrichsen notes that the nonprofit

Damiane Nickles of Closed Loop Farms describes the food raised at The Plant.

has been "going more toward the people" of the neighborhood. In real ways, Plant Chicago is employing a place-based approach to growing social capital on the neighborhood level. "We serve," Hinrichsen continues, "as a convener of people, creating a space where they can have collaborative experiences." This is important in a neighborhood such as Back of the Yards, where traditional supplies of social capital are often in short supply. Once home to a white European population that took advantage of well-paying industrial jobs, the community is now predominantly Latinx. Thirty percent of the neighborhood lives below the poverty, while approximately a third of the population over twenty-five lacks a high school diploma. Close to half the residents between the ages of twenty and twenty-four are unemployed.[15]

It is such realities that Plant Chicago addresses through The Plant. The nonprofit uses the buildings' aquaponic systems and other components to provide free STEM programming for nearby Chicago public schools, including Chavez, Hamline, Sawyer, and Hedges Elementary Schools: close to three thousand students participated in such programs in 2016 alone. As Hinrichsen explains: "We asked of the community, 'What is the need?' We saw that there were no science-based things to go to at all." For nonstudents, Plant Chicago works to create sites of potential economic activity. Through a grant from the USDA, it has been able to start indoor and outdoor farmers' markets on-site. According to Hinrichsen: "Half of our farmers' market vendors come from the neighborhood." The nonprofit has also worked with organizations like the Ellen MacArthur Foundation to bring resources to nearby businesses like the Latinx-owned Back of the Yards Coffee and groups such as the Peace and Education Coalition of the Back of the Yards. "We try to do a lot of work with the neighborhood," Hinrichsen concludes.[16]

Much of this emphasis on working with those nearby is in order actively to work against the possibility of the site becoming an agent of gentrification. Commenting on this possibility, Hinrichsen admits: "We try to keep it completely in mind." And she notes that The Plant works "to have the tenants think about these things as well." She will assist the for-profit firms in finding local suppliers and in hiring from the surrounding community. In this way, The Plant is an important social bridging site for the community. While many gentrifying newcomers economically and socially exchange with more privileged portions of the city exclusively, The Plant is a conduit for interaction with the neighborhood. Local hiring and buying represent one key aspect of that approach, facilitating the neighborhood's capturing of income. And, for local businesses exchanging with The Plant and its tenants, Back of the Yards residents can keep the profits and the related power. The bottom line for Hinrichsen is the desire to create a place that is an economic, social, and cultural asset for the building's neighbors. When told that her work could be seen as operating within the framework of creative placemaking, Hinrichsen sheepishly admits that she is unfamiliar with the concept. Given her background in environmental science and agriculture—she has worked on permaculture farms in Ecuador and Colombia—place for her has always been more about production and meeting basic human needs than unleashing creativity.[17] This approach should serve the organization well as it expands beyond The Plant, building a new headquarters in a nearby derelict firehouse.

But the story of The Plant and its experimentation with revitalizing place starts fourteen years before with John Edel's first project. In the early years of the twenty-first century, Edel was rounding out a decade of building virtual sets for the postproduction firm Post Effects and getting tired of high-tech work. He wanted to reconnect with his industrial design roots. In 2002, he established Bubbly Dynamics, LLC, with the stated goal of "incubating small businesses in formerly vacant, industrial buildings located in disinvested communities."[18]

The company's first project was the creation of the Chicago Sustainable Manufacturing Center (CSMC) in the city's Bridgeport neighborhood. The site that came

The Plant's John Edel experimenting with growing systems at Bubbly Dynamics.

to house the center, a twenty-four-thousand-square-foot, three-story building that previously served as the Lowe Brothers Paint warehouse, was purchased by Edel in 2002. At the time of the sale, the building was derelict and did not even have running water. Five years and some very flexible tenants later, the center was ready to open and ready "to provide an applicable model to the manufacturing sector for ecologically responsible and sustainable urban industrial development." The center's commitment to sustainability can be seen in the way Edel was able to get the site up and running. Not only did he rebuild with salvaged materials and supplies diverted from the waste stream; he relied heavily on volunteer labor and bartered with local suppliers for needed goods. This approach was motivated by a self-consciously DIY ethic honed with the members of the Rat Patrol, an anticonsumer anarchist bicycle club that famously makes "Frankenbikes" from dumpster-dived frames. This collaborative and DIY approach meant fewer external constraints. Edel did not need to repay loans or create programs to satisfy funders. The building and its tenants developed far more slowly than they would have with a traditional development approach, but the freedom allowed organic expansion. These practices carried over into the ecological commitments of early tenants, like Dolan Geiman and Ali Walsh, whose arts practice reworked salvaged materials from to-be-demolished buildings.[19]

Since then, Edel has used this freedom conscientiously to craft an environment where there can be a blending of knowledge- and production-based industries and the color of your collar—white or blue—does not really matter. By 2019, the building housed such tenants as 48 Industries (custom screen printing), Pedal to the People (bike repair), Wan Gerin (bike frame construction), MP Custom Made (furniture making), and Dansch, LLC, an industrial design and engineering consultancy for manufacturing firms in Chicago. Other tenants offer access to the knowledge and materials needed to become makers, including the South Side Hackerspace and Make Chicago, with the latter offering access to space, expertise, and tools as well as courses on beginning woodworking, furniture refinishing, upholstery, and welding. As would be seen later in The Plant, the center is, Edel stresses, all "about creating jobs for other people and maintaining—or trying to rebuild—the industrial base in the core of the inner city, where it should be, especially in these industrial areas which have seen such tremendous job loss."[20]

On the one hand, such urban locations are perfect for economic revitalization efforts because, as Edel points out, they have a large pool of available workers. But they also still have the infrastructure needed for growth, particularly in the form of usable buildings. To Edel, the citywide policy of demolishing buildings at the onset of urban renewal efforts was a great waste. And this policy still informs the mindset of many city officials; as recently as 2015, Chicago mayor Rahm Emmanuel was calling for an additional $35 million to be allocated to the city's budget for building demolition.[21] On the one hand, there is little doubt that the Great Recession led to a glut of abandoned buildings throughout Chicago and that something had to be done with such structures. Yet Edel sees in such efforts a one-size-fits-all approach to addressing abandonment, the policymakers involved considering demolition

the quickest route to improvement. According to Edel: "It's so frustrating to hear them talk about improvements by leveling things. And that's always been the case, for so long—even before the first Mayor [Richard J.] Daley. But I think enough of them are starting to see the light." More specifically, he notes that former Chicago alderman James A. Balcer (Eleventh Ward) came to see the benefit of salvaging derelict buildings and that he managed to convince a large "old-school" Chicago developer to consult with Edel—a conversation that made Edel realize he was now in the big leagues. The long-term sustainability of both the CSMC and The Plant, along with the positive media coverage that both ventures have accumulated over the years, is leading politicians and policymakers to see the value of Edel's adaptive reuse projects.[22]

Yet Edel is also quick to point out that his method also employs a history of the neighborhood that very few tend to privilege or even acknowledge in communities like Bridgeport. The CSMC's website notes how the area faced a significant industrial crisis in the late nineteenth century when the stockyards that the South Side of Chicago had relied on so heavily had begun decentralizing throughout the Midwest and the woods of nearby Wisconsin, the source of raw materials for the community's lumberyards, had begun to wane. According to the Illinois Historic Preservation Agency, the neighborhood faced by the 1890s a series of "economic and geographic pressures," including "the expansion of manufacturing activities, the availability of labor, wage prices, scope and evolution of markets and suppliers, political and social pressures, and the physical geographical constraints of the city," an explanation that fits the postindustrial moment a century later. By the turn of the twentieth century, many feared that the neighborhood's economic future was not looking good.[23]

In response to this turn of events, Frederick Henry Prince, the owner of the Chicago Junction Railway, and A. G. Leonard, the president of the Union Stockyards and Transit Company, oversaw the 1902 planned consolidation of nine smaller railroads into the Central Manufacturing District (CMD). The CMD sought to use its access to such transportation networks as the means to create what the CSMC's website refers to as "the nation's first planned industrial district," one that played up "the incubator function of the CMD": "It was particularly attractive to new businesses, or businesses from out of town wanting to test a Chicago location." The CMD purposefully targeted small manufacturers and offered them things like flexible leasing arrangements or purchase plans and access to a pool of approved contractors for building jobs, the CMD Bank, which provided tenants fair lending terms, and favorable deals with freight carriers.[24]

The building that the CSMC would come to call home was designed by the architect Frank L. Smith in 1911. While some prospective tenants blanched at the multifloor layout (as they did not want to navigate either stairs or an elevator), the building's "compact style," according to the CSMC, "lent itself to the CMD's incubator function." Within the space, upstart firms not able to afford an entire building could lease a smaller unit within the space. Yet what made the building particularly attractive to countless companies was the fact that they would be in close proximity to others engaging in complementary work. Companies quickly flocked to the CMD. By 1915, it was home to close to one hundred businesses, employing, in combination with the nearby stockyards, almost forty thousand people.[25]

It is easy to see why this history proves appealing to the likes of John Edel in the twenty-first century. Here, we see a response to a moment of financial/labor crisis that called for a turn to smaller manufacturing firms and thus allowed for new actors to enter the city's evolving economy. Yet Edel sees in his work a challenge to heal the damage wrought first by this earlier history of industrial expansion and then by the response to the contraction of this expansion, which was almost as degrading to the environment as that initial moment of industrial development. Edel named Bubbly Dynamics, LLC, in honor of Bubbly Creek, the name given to an arm of the nearby Chicago River and popularized by Upton Sinclair in *The Jungle*. The creek collected the drainage of the nearby packinghouses, turning it into "a great open sewer a hundred or two feet wide," one in which "the filth stays there forever and a day." Edel's work with the CSMC and The Plant suggested a new form of environmentally conscious, place-based economic redevelopment. Such a commitment to sustainability is not only not a part of standard urban economic-redevelopment strategies; it is also not usually a part of contemporary placemaking efforts, which tend to prioritize social objectives rather than conceptualizing the interrelationship between the two from the beginning.[26]

These histories that inform Edel's work, much like the histories that inform urban agriculture, are not privileged in contemporary placemaking efforts; indeed, they are often ignored entirely to focus on the postindustrial future, a future in which culture drives a knowledge-based economy that seeks to attract arts-oriented creative types, not those actually manufacturing things. Moreover, to many creative-placemaking advocates, manufacturing—with its history of pollution (a history that Edel has fought hard to overcome)—is best left in the past. Yet others are beginning to draw from them, including Philip Cooley in Detroit.

Cooley is best known for his role in developing Slows Bar BQ, a restaurant that has become a Detroit staple and a media magnet, with locations in the city's Corktown and Midtown neighborhoods as well as a stand in Ford Field and full restaurants elsewhere in the state. But, before all that, Slows opened as a single restaurant on Michigan Avenue in 2005 in the shadow of the now infamous former Michigan Central Station. It is hard to imagine today, but this was before Detroit was selling newspapers as the poster child of the Great Recession or, later, as a member of the country's hippest-cities lists. At the time, Slows' Corktown neighborhood was best known as the place recently abandoned by the Detroit Tigers, and the derelict former stadium was a hole in the neighborhood's heart. Bars and other businesses once driven by baseball fans

The Michigan Avenue block anchored by Slows (*on left*) in 2009, 2012, and 2015.

were closing, and the community's future was once again unclear. Today, Slows' Michigan Avenue block is one of the trendiest in the city, with Detroiters, suburbanites, and tourists stopping not only by Slows but also by slow-drip Astro Coffee, the classic cocktail bar Sugar House, the upscale Mercury Burger Bar, the bicycle shop Metropolis Cycles, the dive bar LJ's Lounge, and another of Cooley's restaurants, Gold Cash Gold. Of the group, only LJ's predated Slows.[27]

Seeking purpose following graduation from Columbia College in Chicago, Cooley decided to move to his father's hometown and work on a project that he had been considering for several years: creating a place that could be the focal point of a community. After living in Chicago and several cities in Europe, he "always felt like a number": "Detroit was that really good blend for me of culture and community." For him, that culture was best represented in the city's eclectic music scene. "When I lived in Europe," he continues, "I heard more Detroit music than anything else, from hip-hop like J Dilla to Jack White to Derrick May. Everyone was talking about Detroit in that sense." The energy that this culture represented, he believed, could be harnessed to address the abandonment that marked Detroit in the early twenty-first century in a way that embraced the inclusive nature of Detroit's music scene. Cooley saw this moment as "a chance for us to create a balanced and sustainable urban landscape."[28]

Building on his family's experience in real estate in Marysville, a small city an hour up I-94, he and his family bought their first building for $45,000 before quickly following up with two more. One building would become Slows, but another would be split into Phil's apartment and his brother's real estate office, an offshoot of the company founded by their grandfather. As Detroit's population continued to fall at this time—the Greater Downtown region (which includes the Corktown neighborhood) experienced a 13 percent population loss between 2000 and 2010—housing stock across the city continued to deterio

rate and disappear. Cooley's decision to create new housing within such an environment at this particular time is worth noting.[29]

Starting a business in a place like Detroit reveals something that is obscured elsewhere. While the politics of opening a new business in many US cities is glossed over to prioritize the quality of the establishment—Is the food good? Are they going to make it?—Detroit is different. Especially in the first decade of the twenty-first century, establishing a new restaurant or a store was big news: not just that someone was establishing a new dining option for residents but that someone was investing in the city, which would mean winners and losers. After all, a business is an agent of capitalism. And Cooley is no stranger to this critique. Among the many concerns levied at him has been his part in the radical change of Corktown's new core. His business was thriving, but was it for the sake of the neighborhood residents or for someone else?

Motivated by such critiques, Cooley began to take on other projects in the neighborhood. In 2009, he started coordinating efforts around the derelict Michigan Central, working with volunteers from area companies and nonprofits to pick up trash, plant native grasses, and the like. This attention to larger-scale efforts undoubtedly informed Cooley's decision to undertake Ponyride, a nonprofit housed in a thirty-three-thousand-square-foot, two-story industrial building in the Corktown neighborhood. Cooley purchased the foreclosed-on property for $100,000 in 2011 as, in part, "a study to see how the foreclosure crisis can have a positive impact on our communities." Ponyride would soon cast itself as "a catalyst for deploying social capital to a diverse group of artists, creative entrepreneurs and makers who are committed to working together to make communities in Detroit sustainable." It is a shared work space with an emphasis on light manufacturing and a stated commitment to the sharing of tools, ideas, and networks among tenants. As of 2018, tenants included Mitch Creative Digital Marketing, the

Volunteers organized by Phil Cooley plant grasses in front of Michigan Central.

Lip Bar (a cosmetics manufacturer), Heritage Works (a dance studio), and Smith Shop (a metalworking studio), among over forty others. As a testament to the success of the building, there is a waiting list of businesses wanting to move in. Those unable to secure space within the building, however, are able to participate in the Ponyride Market Summer Series, an outdoor market that allows vendor space for many locally based businesses. The market began as a stand-alone event in 2011 and became a monthly series in 2017. In addition to providing more opportunities for small businesses, it opens up the otherwise inwardly focused workplace and provides an opportunity for tenants to reach new audiences.[30]

Of course, Ponyride is not the first such shared work space. What is noteworthy about the building, however, is the thought that has gone into the model that informs the space. The site employs six parameters to structure the relationships between building, tenant, and community: lower-than-market-rate rent for space, the build out of space by tenants and volunteers, diversity of actors (skill set, background, race, etc.), education and community outreach by tenants, deconstruction and reuse of reclaimed and salvaged materials, and activation of an underutilized and discarded building. This means that the ability to charge tenants $0.50–$0.65 per square foot—or one-quarter of market rate in the neighborhood—is predicated on the fact that individuals and organizations working in Ponyride have put roughly $200,000 worth of volunteer hours and materials into what was previously an all-but-vacant building (which allowed it to be purchased in the first place at such a low cost). Here, the line between DIY and neoliberalism may be razor thin, but the arrangement allowed those without capital to find an initial home for their operations, a reality that led Ponyride to attract a racially and economically diverse group of tenants. Such diversity, along with the requirement that tenants must share their expertise with the community by providing at least six hours of free classes per month, allows Ponyride to maintain a positive relationship with the surrounding community while at the same time attracting

potential new tenants. And the space continues to pay attention to the need to revisit these parameters and the relationships they are meant to foster. In September 2017, for example, Ponyride became the first Detroit coworking space to offer on-site child care.[31]

But what is perhaps most noteworthy is the way manufacturers use the facility as a place to begin their business and then move on to another space. Here, those at Ponyride seem to be mimicking the growth strategy of earlier producers in Detroit and elsewhere, whether it be Packard moving out of a small space into its grandiose auto-manufacturing plant in 1903 or Hewlett-Packard legendarily outgrowing the Palo Alto garage in which it made its earliest products in the 1930s. The for-profit Detroit Denim—a jean manufacturer that set up shop in the Ponyride building in 2011—moved to a larger production space and retail shop on the city's East Side in 2016. Even nonprofit organizations such as the Empowerment Plan have used the building as a sort of incubator space, giving them the physical room and tools needed to begin and grow quickly. Since 2012, the Empowerment Plan has trained, through a two-year "stepping-stone" employment program, forty-five homeless individuals to manufacture over twenty-five thousand coats that transform into sleeping bags for housing insecure individuals. Importantly, all the Empowerment Plan's employees have gone on to find permanent housing.[32]

The success of this first iteration of the Empowerment Plan compelled the organization to find a larger base of operations. In late 2017, the nonprofit announced that it was leaving Ponyride for its own twenty-one-thousand-square-foot building in the city's West Village neighborhood. This new site would give it classroom space to run additional financial literacy programs, GED training, and other workshops for the organization's employees. It would also give it the space to expand Maxwell, its new for-profit coat-manufacturing business. "The revenue earned from [Maxwell's] sales goes back into Empowerment Plan," explains Empowerment Plan founder and CEO Veronicka Scott.[33]

Workers sew garments at the Empowerment Plant, one of the
Ponyride tenants.

Despite such high-profile firms leaving the building, Ponyride still found itself turning away deserving potential tenants on a regular basis. In part to deal with such concerns, in early 2018 it announced that it would be moving to a larger building in the New Center neighborhood, sharing space with the alternative arts group Make Art Work and the education and recycling center Recycle Here! Yet this decision to move was rooted in more than just spatial concerns. As in cities like Miami experiencing population growth, the presence of a burgeoning hospitality industry has lured other industry to the area. The success of the commercial district now surrounding Slows and Ponyride attracted others to the Corktown neighborhood: two research facilities for composite and lightweight materials opened in 2017, and, starting in February 2018, Ford was renovating a forty-five-thousand-square-foot formerly vacant industrial building into a facility meant to develop driverless cars and electric vehicles. The fact that this dynamic was unfolding in Rust Belt Detroit highlights just how powerful the Ponyride project became over time. Moreover, many observers feel that the most recent plan to redevelop Michigan Central—this time another Ford facility—might actually stick. Such development has led to a rapid rise in real estate values in the community surrounding Ponyride. Slows and Ponyride have not been priced out of the neighborhood, but Cooley is aware of how his socially engaged, place-based work has produced disparate economic consequences. As he notes: "As rents keep increasing, there are folks that keep getting left behind or pushed out. There's a lot of folks that can't fit into the traditional system that are capable of being innovative, creating jobs and creating the economy we need to be successful." The hope is that the New Center community will provide such necessary affordable spaces once again and that they can stay that way. However, increasing private and public investment in the neighborhood, including the new QLine streetcar that connects New Center to downtown, suggests that that dream may be difficult to realize.[34]

Other urban businesses throughout the country are also wrestling with the relationship between community development and displacement. Elsewhere in Detroit, Ann Perrault and Jackie Victor opened Avalon International Breads in 1997 with the aim of following the Buddhist principle of right livelihood, which has meant practicing environmentalism and community development and providing their employees with fair wages and benefits. The next block over, small businesses such as City Bird (a boutique that carries Detroit-themed housewares, accessories, and paper goods) and the now-shuttered Bureau of Urban Living (a self-professed "urban general store") started organizing events like the Canfield Street Market, the first of which was held in June 2010. This event, which gave artists and small businesses the opportunity to display and sell their wares, was presented in conjunction with the radical social justice organization United States Social Forum, and a portion of the proceeds went to the East Michigan Environmental Action Council. Canfield Street Market was given much publicity by fledging sites like iamyoungdetroit.com, a site run by the artist Margarita Barry meant to highlight inspiring Detroit residents under the age of forty. Barry herself opened 71 POP, a retail gallery showcasing the work of seventy-one emerging artists and designers, in a previously abandoned Detroit building in the summer of 2011.[35]

One result of all these efforts is that the Midtown neighborhood these socially minded businesses helped create is one of the most prosperous in the city. And, as is seen in Corktown, even in a place with as much open space as Detroit, development and displacement are often interconnected. Moreover, all this is occurring as others seek to remake Detroit on a grand economic scale, often using the ideas and language of creative placemaking to justify their efforts. As we noted earlier, the Fortune 500 leader and Quicken Loans founder Dan Gilbert has partnered with PPS and even gave an address at the PPS inaugural Placemaking Leadership Council meeting in 2013, held in Detroit. Often, such larger-scale redevelopment

The interior of Avalon International Breads.

efforts use the sheen of such grassroots efforts to provide authenticity—and access to a premade customer base. It is no accident that, in 2013, a new Whole Foods grocery store (the opening of which was championed by Gilbert and other city leaders) set up shop just a ten-minute walk from Avalon International Breads.[36]

While Avalon and Whole Foods are now partners, with the bread company's products available in all Michigan Whole Foods locations, when Whole Foods opened up, Avalon experienced a $350,000 drop in revenue over the next twelve months. Moreover, now across the street from City Bird is the flagship store of Shinola, a lifestyle brand started by the founder of the global watch company Fossil Group. Shinola has positioned itself as a Detroit-based heritage brand, thanks to buying the rights to the Shinola brand name and the heavy promotion of the company's Detroit factory, where it assembles watch parts imported from abroad. It was just the first of other high-priced, high-profile stores that moved into the area, and that was even before the new $1 billion Little Caesars Arena and surrounding district broke ground.[37]

Such developments suggest that we should be wary of the dynamic in operation between community development and community displacement. Yet the example of Detroit also begins to suggest how a much place-based understanding of redevelopment is influencing a more mainstream conception of both economics and place-making. Here, the relationship between Whole Foods and Avalon International Breads is indicative of a broader phenomenon. Commenting on this relationship, the Whole Foods representative Christine Sturch notes: "We decided rather than compete with all these people who were so passionate about food, that we would partner with them. This was a new model for us." In addition to working with Avalon, Whole Foods also partners with Eastern Market farmers to get their produce into the grocery store.[38]

Other companies have followed suit. In 2016, Nike opened one of its Community Stores—the Community Stores being a series of stores meant to address disinvest-ment and unemployment directly in underserved neighborhoods—in Detroit. One of the stated goals of these stores is a commitment to hiring from within the city; in Detroit, 80 percent of the retail team lives within a five-mile radius of the store. Through the Community Store, Nike also contributes $40,000 in annual grants to local nonprofits. Even much-maligned Shinola has shown a commitment to employing Detroit residents, going from ten workers in 2011 to over two hundred employees at its Detroit facilities by 2016. In 2015, the company also created its REACH campaign, which has partnered with such organizations as the CityPak Project—which provide backpacks filled with supplies for Detroit's homeless population—to deliver services to the city's most vulnerable residents. Such cases are undoubtedly complicated as private companies cast themselves as both job creators and service providers. It is still too early to tell whether such programs will lead to increased gentrification in Detroit; they should be assessed over time as they seek to establish roots throughout the city's neighborhoods.[39]

Survival of the Smallest? Entrepreneurialism at the Neighborhood Level

Such activity does not mean that those trying to effect change at a more grassroots level have gone away; others in Detroit have little interest in working with larger corporations like Nike or even Shinola. A push for smaller-scale, community-based economic redevelopment, for example, can be seen in the efforts of Friends of Detroit and Tri County on the East Side of Detroit, seemingly a world away from Midtown's economic boom. This nonprofit organization has used job training and creation as mechanisms to cope with the loss of employment related to the recent financial upheaval: the group's promotional material reports that the region "lost an estimated 150,000 manufacturing jobs in the last few years." Yet this history of crisis goes back even further for the group, with the transition "from muscles to technology" in the late 1960s

The Hope District's Conflict Resolution Garden.

setting the stage for a process of deindustrialization that has gone on for fifty years. For Friends of Detroit, it was frustration at such a transition that led to the 1967 Detroit rebellion, a racially motivated episode between the city's African American population and police that led to forty-two deaths and over seven thousand arrests. Rather than casting the rebellion as a nadir in Detroit history—or the end of postwar affluence in the city—Friends of Detroit instead posits that it highlighted that African Americans had never truly shared the benefits of Detroit's rapid growth throughout the twentieth century. The events of that hot summer were, then, not an end point but a starting point: "The 1967 Rebellion had to happen for the hope of the grassroots to be unleashed." For Friends of Detroit, the goal is to keep this commitment to grassroots civil rights work, this flame, alit in the new millennium. "The flame of '67," the group proclaims, "still burns for peace, social, political and economic empowerment." This means that any attempt to address this climate must go beyond the cosmetic and focus on the structural. For Friends of Detroit and Tri County, the answer to this is "to create a skilled local work force that can rebuild their neglected communities through new home construction and enhanced economic development."[40]

The centerpiece of the group's efforts is the Hope District, "a manifestation of that flame." The Hope District is an approach to grassroots economic development meant to engage the residents on the East Side of Detroit, where the poverty rate can reach 50 percent, directly. Within this region, Club Technology, a twenty-three-thousand-square-foot building funded through a combination of public and private sector funding, exists as the headquarters of Friends of Detroit and Tri-County. The building contains fifteen classrooms, a full-service commercial kitchen, a number of offices, and a large multipurpose room, all designed to provide resources for neighborhood residents. The organization is supported in part by government social service funds designed to employ hard-to-place individuals, but it does much more than that. It offers weekly arts and crafts classes and workshops for the community in the building, which is also available for rent to the general public. Somewhat less technologically driven, community gardens and orchards are strategically placed throughout the district, meant to provide healthy produce "to feed the people of the community."[41]

Yet the vision for Club Technology is that it be more than simply a community center. The organization is working with technical consultants from the city of Detroit to turn this area into a self-described "grassroots cyber district" and providing training in Internet-based skills for area residents, creating new relationships and opportunities in both the online and the physical worlds. The Hope District thus envisions itself as part of Detroit's broader DOCTOR Project, a program meant to reinvigorate vacant and underutilized commercial strips throughout the city. Friends of Detroit and Tri-County hope that the creation of such a center would then lead others to revitalize surrounding housing stock, undertake facade improvements, and tackle other projects—all of which would provide potential sources of employment.[42]

To Mike Wimberley, who leads the Friends of Detroit and Tri County on his own after the passing of the or-

ganization's founder—his mother, Lily Wimberley—such efforts are needed to push the economy, in Detroit and elsewhere, in new directions. As he notes: "The ripple effect of globalization and our push to make finance the end-all, be-all of entrepreneurship in our economy is that, because financiers and banks are calling the shots, people are finding themselves displaced. In this environment it's important we help people manifest their destiny." A crucial component of such a strategy involves helping community members find ways they can work and make a living in such a changed economic climate. "Everybody has skills," Wimberley continues. "If they haven't found them, we have to work with them to find them. Then, as an agency, we must create an infrastructure that helps each person ramp up their skills until we can figure out how to monetize them." Here, the Hope District has several strategies for monetizing skills. One comes from a book rarely cited by mainstream placemakers: *Building Communities from the Inside Out: A Path toward Finding and Mobilizing a Community's Assets* by John P. Kretzmann and John L. McKnight.[43]

For Kretzmann and McKnight, most experts concerned with urban redevelopment dwell on a community's deficiencies and problems and then seek additional outside expertise to address them. This is the approach that many cities take when they begin a creative-placemaking campaign: call in an outside expert like Richard Florida to find out how the recruitment of the creative class could address that city's problems. A second path, however, is based on "asset-based community development." This model seeks to discover and strengthen capacities already present in the community and then use such discoveries and the relationship-building work that brought about such capacities in the first place to develop new assets. These assets are then physically mapped out, allowing residents to see what is available to them and that they are "part of the action, not as clients or recipients of aid, but as full contributors to the community-building process." It is clear that this is the work Wimber-

Hope District participants play games in the Club Technology building.

ley is doing—drawing from and training local talent—as he leads a group whose motto is "There Are Enough of Us to Help the Rest of Us."[44]

There is little doubt that Wimberley was also connected to the life and work of Grace Lee Boggs, the legendary Detroit activist who passed away at the age of one hundred in 2015. Not only is the Hope District mere blocks from the Boggs Center, Boggs's former home and a decades-long center for her activism with her husband James, but Wimberley also served on the Boggs Center board and was a favorite example of Boggs's concept of "(r)evolutionary" action in Detroit. Evidence of their collaboration

through reading groups and activism abounds, including in Boggs's 2011 *The Next American Revolution: Sustainable Activism for the Twenty-First Century*. In a chapter titled "Detroit, Place and Space to Begin Anew," Boggs finds that "Detroit is a city of Hope rather than a city of Despair." The ways capitalism has failed the city provide an opportunity to build "a new kind of economy from the ground up."[45]

Creating this new model is the theoretical and practical context of the Hope District. It is not just that Wimberley is focused on the communities with which he works; he is also concerned about how important funding is for

Grace Lee Boggs celebrates with the actor Danny Glover during her ninety-fourth birthday party.

such grassroots projects across the city. To Wimberley, a reliance on grants and other sorts of philanthropic gifts is simply not sustainable, for both financial and ideological reasons. "On one hand," he explains, "we have to recognize that philanthropic money also comes with strings attached." Moreover, grantmaking bodies such as the NEA and others "are often adherents to the status quo": "And, as Louis Farrakhan said, the status quo for those on the bottom is like asking us to stay in the crack house." Ultimately, then, Wimberley concludes: "We have to have the flexibility to have autonomy."[46]

This autonomy allows Wimberley to act creatively. But that autonomy can be maintained only if he is able to pay workers. As he stresses: "We have to figure out how to monetize—a term some people may think crude—to monetize what we do to be able to pay for the lights, the gas, and ultimately to be able to pay competitive wages so our institutional knowledge doesn't walk away to work for a bank or somewhere else." The lesson is that organizations have to be entrepreneurial and expand opportunities for their volunteers and employees so that they will stay with the organizations as their life courses change. They may stay while they are young, but, if there are not more opportunities as they age, what happens as they get older and have children of their own?[47]

The Hope District publicly expresses this desire for

sustainable community development throughout the life course through signs proclaiming messages like: "Start your own healthy living business now.... Licensed commercial kitchen available." The emphasis on food is not just an afterthought. Cooking and growing are among the community's self-identified strengths, and the Hope District supports community gardens. Seeking the highest potential from these assets, Wimberley knows that a processed food product can be sold for higher profit margins than could be had with raw fruits and vegetables and that more profit means more money to invest in the community. The group had early success selling sweet potato pies, and the current standout is Detroit Friends Potato Chips. Wimberley developed the idea to sell organic potato chips in 2012 and then refined it through FoodLab Detroit, a nonprofit that helps Detroiters develop food-based businesses with a social mission. The initiative has grown to employ a dozen workers who do everything from grow, cook, and distribute the potatoes through local grocery stores and markets. Mail-order purchases are also surprisingly large thanks to *O* magazine, in which they were featured as one of Oprah's favorite things of 2016.[48]

The project illustrates how Detroit seems poised to provide an environment where these kinds of activities can happen. "Because the city is in the position it's in," Wimberley explains, "it's open to experimentation, especially when it's used with unused land and in a community that's on the margins. So the city has been as accommodating as a broke city can be in terms of what we're doing." But it is not just the city's inability to monitor development or its openness to experimentation that allows for this; Wimberley says that there is also a culture and institutional flexibility to work in places like Detroit. "Rem Koolhaus," he concludes, "has it right: visionaries in large cities with developed construction industries and the groups that oppose them are straitjacketed, but places in like Detroit, and those places in Africa, the sky's the limit because people don't have to worry about the cultural constraints and backlash to their projects." The private sector has started to warm to this idea as well. The Detroit

Creative Corridor Center, which provides infrastructure, consulting, and other resources for start-up businesses in film, fashion, architecture, and digital media, has been in operation since 2008, and Green Garage Detroit, an incubator for environmentally friendly companies, has been thriving since it opened in August 2011. Such examples provide evidence that this model of place-driven community development can succeed without irreparably transforming the nature and constitution of the communities from which they are organized.[49]

Urban—and Spiritual?—Renewal: Religiosity and the Rethinking of Place, Work, and Economy in the American City

The entrepreneurs we have examined so far are committed both to a place-based concept of urban redevelopment and to an understanding of economic exchange that often challenges the status quo. Yet there is another element to this alternative approach, one rooted in religiosity and spirituality, that has received virtually no coverage by the likes of such mainstream publications as the *New York Times*, *Forbes*, or the *Wall Street Journal*. For the Detroit-based United Methodist pastor Bill Wylie-Kellermann, the role of spirituality in the rebirth of the city becomes apparent when attempting to answer a very specific question: "What if the history of suffering in Detroit is a spiritual resource waiting to be unleashed?"[50]

In attempting to answer his own question, Wylie-Kellermann recasts the rise and fall of Detroit, a history that has played out in countless US cities and is usually ignored or seen as the problem to overcome by contemporary placemaking efforts, as a reservoir of resilience, the raw material of resurrection. For Wylie-Kellermann, this history matters. On the one hand, Detroit residents have crafted strategies to deal with the loss of jobs, services, and income even before those problems were associated with the more recent mobility of capital. But this history also exacts a toll on those in Detroit: it is almost as if this collective past of anguish has been welling up for

Pastor Bill Wylie-Kellermann in his church.

years and the current financial collapse finally broke down the dam. Now let loose, this pain can become an asset, a means of reshaping the city. Within these longer stories of deindustrialization, job loss, and abandonment, "hope begins as grief."[51]

Fueling this belief in the possibility of conversion is the presence of what other observers see as remnants of a history of declension: a landscape of vacancy. By 2009, close to 30 percent of Detroit—nearly forty square miles— was vacant land. As politicians struggled with a response to this dereliction and its potential effects on job development, some saw potential. That year, Wylie-Kellermann watched eleven thousand urban gardeners connect to

Detroit's Garden Resources Network's three farms, over three hundred school and community gardens, and nearly six hundred family plots. One could view the explosion of such garden spaces through the lens of religion, as faiths across time and space have held such sites as sacred. Yet Wylie-Kellermann is also interested in the way such places empower city residents. These local entrepreneurs did not wait for the go-ahead from the government or private-sector actors; they saw an available resource, and they utilized it.[52]

When pressed in the fall of 2010 to expand on these ambitious ideas, Wylie-Kellermann choose to elaborate on the concept of a spiritual economy. The impact of the financial crisis highlighted the need to think "less in terms of hoarding and consolidation of capital and commodities, and more about gift, release, and grace." Cities like Detroit now have the opportunity to "build a new economy from the bottom up," and this new economy, distinct from the "finance-based Wall Street economy," must instead be evaluated "in terms of how it builds literal community."[53] Wylie-Kellermann is indicative of how many activists— particularly in Detroit—have begun to blend spirituality, place, and economic justice. Recalling Martin Luther King Jr.'s late-in-life call for a more humane approach to labor, Grace Lee Boggs noted that the early twenty-first century needed "to relocate God": "We need a new philosophy to provide a framework for our economic and political decisions, a philosophy with a more unified concept of the spiritual and the material, a more immanent concept of the divine and a more sacred concept of Nature." And the Central United Methodist Church was instrumental in organizing the 2010 United States Social Forum, an event held in Detroit that focused on the idea that "another world is possible," particularly when it comes to matters of economic exchange, and attracted more than twenty thousand radical activists.[54]

This other world has little room for the type of place-making exemplified by the likes of Dan Gilbert. On June 18, 2018, as part of the Poor People's Campaign's Forty Days

A sign for a long-shuttered restaurant stands over a vacant lot in Cleveland.

City crews mow derelict lots on Detroit's Southwest Side.

of Action, Wylie-Kellermann was arrested for blocking the QLine, a 3.3-mile streetcar service whose moniker comes from Gilbert's Quicken Loans company (Gilbert bought naming rights to the line in 2016). Wylie-Kellermann considers the QLine "emblematic of development priorities, downsizing the city's foot print at the expense of neighborhoods and their destruction." Such projects are privileging the creative class, which has little to offer the city's indigenous population. In fact, such work actively takes resources away from communities that exist outside this redevelopment narrative. Wylie-Kellermann is quick to note that public-sector money was found to help fund the construction of the QLine "while [the] city is cutting

bus service to Black neighborhoods," a move that has profound economic repercussions for those neighborhoods' residents. So, while, as of early 2019, Wylie-Kellermann still believes that the "grassroots, alternative, entrepreneurial economy is alive and well," he is quick to add that it "exists partly in resistance to the dominant narrative."[55]

But what might such alternative economic activity look like? And how might it come to provide the resources that more mainstream understandings of both redevelopment and placemaking seem to take from lower-income communities of color? A group working on both the South and the North sides of Milwaukee can provide us with a more grounded example of how place, economy, exchange, and labor are being rethought in a large American city. Traveling from downtown Milwaukee to the city's Near South Side is like taking part in some sort of strange archaeological exercise: everywhere one looks there are earlier attempts at dealing with moments of profound economic crisis. Driving through the center of the city, one sees such structures as the Shops of Grand Avenue, an indoor mall that was erected in 1982 to address the flight of both (white) people and businesses from the inner city to Milwaukee's booming suburbs. The 2007 economic collapse dramatically increased the vacancy rate at the mall, and many within the city are beginning to doubt its long-term viability. Crossing the Sixth Street Viaduct, the Potawatomi Casino and the Harley Davidson Museum both stand as examples of Milwaukee's more recent strategy to rebrand the city as a tourist destination and reclaim the Menomonee valley, ground zero for the disaster that deindustrialization brought to the city. Ironically, three months before the museum opened, Harley Davidson announced that it was cutting over seven hundred jobs and idling its production plants because of the financial crisis.[56]

In the South Side's Walker's Point neighborhood, rehabbed industrial buildings came to house the offices of postindustrial finance, insurance, and real estate companies like American International Group, Inc., health care providers like Aurora Health Care, and high-end apartment complexes. Yet in between these landmarks of an economy that mostly passed Milwaukee by are legions of abandoned buildings, ghosts of the industries, including tanneries, breweries, and other manufacturing plants, that once drove economic development in the city.

Bill Wylie-Kellermann is not the only person to survey such a landscape of abandonment and see the opportunity for both urban and spiritual renewal. In Milwaukee, the environmental consultant and founder of Caleb Acquisitions Michael Frede shares the belief that the Great Recession unleashed a certain spiritual energy, one that can remake both the urban environment and the ways we think about jobs, labor, and development. As a devel-

oper, Frede has made it a point to purchase and reclaim land that is blighted. The decision to purchase such properties is motivated by a higher calling. "I believe that God created everybody for a purpose, an individual purpose," explains Frede. "And I also believe that God created the earth for a purpose and land for a purpose.... Land varies dramatically, and land should be productive. This doesn't have to mean that every square inch of land is developed. Some land's purpose is for beauty, to be artwork. Other land was created to provide resources, natural resources. Other land was created to sustain people and what they are supposed to be doing with their lives." This religiously directed vision drives Frede to work "to restore land to productive use": "There's a lot of land out there sitting vacant, unused, and is contaminated for whatever reason. I feel it is a part of my work to help some of these properties be restored." God has, according to Frede, a vision for every parcel of land, just as he does for every person.[57]

Down the street from a shuttered AIG office (the Great Recession hit the postindustrial economy hard in Milwaukee) is one of Frede's redevelopment projects and its tenant, Community Warehouse. Located in a building that once housed the Blackhawk Tannery, Community Warehouse opened its doors in 2005. Overshadowed by a highway overpass, and tucked behind a gas station, the building has an unadorned facade that hides the remarkable efforts inside. The former tannery has been transformed into a sort of DIY Home Depot, with shelves on shelves stocked with new home- and facility-improvement materials. From windows to doors to paint and even bathtubs, Community Warehouse has it in stock.[58]

There is a specific customer base for such an inventory: Community Warehouse began as a membership organization, with members consisting of neighborhood residents, other nonprofit organizations, and property owners who live and work within a zone marked for redevelopment by Milwaukee's Community Development Grants Association. Members—who come from some of the city's most impoverished neighborhoods—can purchase these home-improvement materials at a price

A Community Warehouse employee at work in the organization's original location.

approximately 75 percent off the original retail value. This once-vacant tannery has again become a focal point for labor and production as the group helps its predominantly Latinx clientele rebuild their homes and their communities. In its first two and a half years in operation, Community Warehouse served 926 households and 67 nonprofit organizations, conducted more than ten thousand sales transactions, collected $2.9 million in material donations at retail value, and made $572,000 in sales (the equivalent of nearly $2 million in retail value).[59]

One crucial aspect of Community Warehouse is the way its work is infused with the idea of religious service: it is meant to serve not only the community's physical needs but, at the same time, also its spiritual needs. Prominent on the organization's homepage is a passage from Matthew 7:24 ("The wise man builds his house upon the Rock"), and those associated with the group stress that they firmly believe that they are doing the work of Christ through their endeavors. By cleaning up and rehabilitating aging industrial spaces, and by allowing economically vulnerable city residents to fix up their homes, the group believes that it is allowing such properties to be "born again."[60]

Here, religion seems to be both a common ground between service provider and consumer and a new vision of the work related to urban redevelopment, one in which, with the assistance of a religiously affiliated organization, the individual leads the way and often performs duties once carried out solely by the state. Such a trend can now, as we will see, be seen in cities across the country as many city dwellers are looking elsewhere for goods and services—and even jobs—once provided by government.

Perhaps most importantly, the group also makes it a point to hire from the communities it serves. Here, Community Warehouse provides jobs to those who may often have trouble finding employment in times of financial crisis, including recent immigrants and convicted felons. The group's former floor supervisor, Jacob Maclin, had been incarcerated multiple times for dealing drugs, and the former executive director George Bogdanovich sees Community Warehouse as being a place that offers such individuals a second chance. For Maclin, who obtained his position through an initiative initiated by the Milwaukee County district attorney's office to get nonviolent offenders back into the workforce, this job provides spiritual as well as material benefits. He notes: "[Community Warehouse has] given me stability through a steady job, a steady home away from home." But it has given him more than that: "[It] also gives us all a place to worship, a place to cherish, a place to grow spiritually as well as emotionally." When asked what his job at the organization is, Rodrigo Cisneros, the group's customer service director, explains that it is "to help people improve their houses, to help give them better lives, and to show them the love of Christ."[61]

In October 2018, Community Warehouse opened a second facility, with Maclin as store manager, on Milwaukee's predominantly African American North Side, in a building that once housed a grocery store (its immediate previous use was as a thrift store). "The community," notes the current Community Warehouse CEO Nick Rinnger, "has welcomed us in." The commitment to Christianity is still a part of the organization's mission. "That's a huge component for me," explains Rinnger, who has a master's degree from Dallas Theological Seminary. The North Side branch seeks to use its site to grow more employment

opportunities, particularly through state-sanctioned programs for formerly incarcerated individuals. Community Warehouse continues to work with the district attorney's office, but over the past five years it has also come to form partnerships with the Milwaukee County House of Correction and federal court reentry programs. Often, according to Rinnger, the appeal from such institutions is remarkably similar: "We hear you're doing a lot of hands-on mentoring. Will you work with these guys?" Commenting in October 2018, Rinnger reports that, of the sixty people who went through Community Warehouse's reentry program over the previous two years, 80 percent found employment, and only one person returned to prison.[62]

All the rehabilitation work on the new store was done by such individuals, who were trained in-house by Community Warehouse staff to do such things as install drywall and assemble cabinetry. One floor of the new space is dedicated to classrooms and workshops as in May 2018 the organization started an eighteen-month mentoring and training program called Partners in Hope. By early 2020, over sixty individuals had graduated from the program. Some of the graduates have found jobs at Community Warehouse, but the nonprofit has positions for only approximately forty-five people. "We can only hire so many guys," notes Rinnger. "There's a bottleneck here when it comes to employment opportunities." The training program will work to find graduates jobs outside Community Warehouse. The organization is also planning to start what Rinnger describes as a "boutique staffing firm," one that will work to find firms willing to hire formerly incarcerated people. For Rinnger, providing such new avenues for potential employment is a true manifestation of his spirituality, one that has the real-world impact of connecting returning citizens with potential employers, two groups that would otherwise have little chance to meet one another.[63]

Back in Detroit, Amy Kaherl, the founder and director of Detroit SOUP and the holder of a master's degree in theology from Fuller Theological Seminary, is allowing for such ideas to inform work being done in multiple

Two Community Warehouse employees lay new tile at the group's new location.

locations across the city. Since February 2010, Detroit SOUP has striven to "promote community-based development through crowdfunding, creativity, collaboration, collaboration, democracy, trust and fun." It is a "microgranting" dinner: attendees pay $5.00 for a meal (the "soup," if you will) and listen to four presenters looking for funding for a particular project. Each presenter has four minutes to share their idea and answer four questions from the audience. There is then time for discussion, after which attendees vote for their favorite presenter— and that individual goes home with all the money raised that evening. As of 2019, Detroit SOUP had hosted more than 175 dinners attended by more than twenty thousand people. Over one hundred different projects have been awarded a total of $147,335, with the bulk of projects funded falling in the category of entrepreneurism. Past winners have included the community-based coffee shop Always Brewing Detroit, the Brightmoor Community Kitchen (a small food business incubator), and the aforementioned Empowerment Plan. These are places, it should be stressed, where people can create, labor, and come together.[64]

On the one hand, these events are seen as places where financial exchange takes place. "The money is the easy way to understand the dinner," explains Kaherl. "There is an economic activity taking place." Yet, in a city where the likes of ubercapitalists like Dan Gilbert garner headlines, Kaherl sees in SOUP a sort of anti–venture capital development ethos. At such dinners, "anyone can sit at the table" in a way that "challenges the idea that you have to have money to make something happen." To a casual observer, Kaherl seems to be calling for a sort of democratization of the funding of placemaking projects, one in which entrepreneurs do not have to rely on highly competitive, big-ticket grants from the likes of the NEA and other organizations. But she also sees SOUP events as a place where democracy broadly conceived can be both exercised and strengthened. "Democracy," Kaherl

stresses, "has to be continually practiced." As more of our relationships and methods of exchange become digitized, "we need to find spaces of actual connection." Moreover: "With the rise of a Republicanism that values individuals, public spaces are gone." SOUP dinners provide such valuable spaces and create a safe public sphere for those— namely, women and people of color—who are often left out of such discussions. In her final analysis, Kaherl concludes: "Money is the least interesting thing exchanged at the dinner." Instead, in language extending the promise of social capital, she notes that she prizes the "intimacy of interactions" that these dinners create and that these interactions ultimately allow for what she terms "human exchange."[65]

Considering her background in theology, such language suggests that Kaherl sees these dinners as nothing less than sacred spaces. The almost reverent tone that takes root during the conversations at these dinners compels her to suggest that these events provide a way "to use the good parts of the church" in the secular world. On a personal level, these SOUP dinners have seemingly reinvigorated her sense of spirituality, allowing her a space where "practicing your ethics" can happen and have a positive impact on the world around you. The money that Detroit SOUP provides entrepreneurial start-ups is important, but Kaherl's involvement with the organization has also led her to reconsider such questions as, "What is love? What is place?" There is little doubt that others feel the same way, as is evident from the spread of the SOUP model: there are now more than 150 such groups across the globe.[66]

By late 2017, the possibility of a global network of SOUP groups led Kaherl into a partnership with the Detroit-based nonprofit organization Build Institute, a self-described "idea activator and small business accelerator focused on grassroots community entrepreneurship." This partnership provides classes and mentorship that give SOUP winners access to an infrastructure of tools that will help make their projects sustainable and profitable over the long run. At the same time, the partnership frees Kaherl from many of the day-to-day organizational tasks that go along with the dinners, allowing her to "build an online—so different from SOUP—hub where people around the world can talk with one another as well as have access to tools and resources" that would provide participants the ability to complete their projects. This would create a sense of "group ownership" within the SOUP model, one that moves the conversation "away from [Kaherl herself] as founder" toward individuals feeling empowered "to make something like this happen in their own community": "Less of a place to have to ask permission and more of a place to grant yourself permission to do it."[67]

Conclusion: Placemaking Is More Than a Conduit

While top-down placemaking often envisions superficial activities as the catalyst for economic development through changing the identity of a place or through facilitating interactions that lead to more substantive changes, the efforts related in this chapter demonstrate how placemaking activities are more than a conduit. Placemaking can simultaneously establish community identity and social capital *and* do holistic community development—with an emphasis on working with residents already in such communities. Moreover, placemakers like John Edel understand that their work is not necessarily new. Instead, they draw from histories and theories that link place-based action to the full range of community change. Doing so means paying attention to the specific history of a place and evaluating its position in a rapidly changing city. This means assessing the broader social, political, and economic landscapes to determine which approaches will be possible, often with few resources of one's own, as well as asking questions like, "What is needed here?" "Can I get funding?" and "Will building inspectors shut

us down or help us out?" After all, when outside organizations operate under the assumption that profit is the number one priority or that groups should fit into specific boxes (e.g., developer or charity), it is not only difficult to be recognized for the scope and scale of work; it is difficult to work at all.

Many groups like Bubbly Dynamics and Community Warehouse seem unambiguously successful in their missions to provide equitable, place-based growth. Yet their success has put them into even deeper conversations—and partnerships—with such state actors as politicians, policymakers, and representatives of the criminal justice system. After all, Edel and his Plant have secured $1.5 million in grant funds from the Illinois Department of Commerce and Economic Opportunity, while Community Warehouse has taken advantage of close relationships with such bodies as the Milwaukee County House of Correction. What will it mean for the future of such efforts and others like them as their methods come to be embraced by the state?

At the same time, the spectacular success of Phil Cooley's Slows and Ponyride projects has played a much more complicated role in the development of Detroit's once-sleepy Corktown neighborhood. Likewise, a few miles east in Midtown, the United States Social Forum's partnership with community-oriented businesses certainly raised the profile and potential of an area now known as one of the most expensive in the city. In a place with as much derelict and abandoned property as Detroit, it is easy to say that any development is good, but, as even Cooley contemplates moving Ponyride out of his neighborhood to make way for the likes of Ford and other potential manufacturers of autonomous vehicles, the challenges of such changes are clear.

Still farther east, Lily and Mike Wimberley's Hope District provides perhaps the most robust solution by asking a community-based question, What resources do we have? As we consider how we fully transform places, the Wimberleys remind us that rooting redevelopment in the community may provide the opportunity to do work that addresses rather than reinforces existing inequalities. The message is echoed by Amy Kaherl's SOUP and its transition to developing a center focusing on grassroots community entrepreneurship of a type that focuses on the actual redistribution of wealth and economic opportunity. Is this a path forward?

In the next chapter, we will show how some of these very conflicts between development and community coherence and autonomy and institutional support play out in urban artistic interventions. We will also emphasize the particular role that the arts can play in shaping these fuller conceptions of placemaking and production. In many ways, art—as positioned by mainstream contemporary placemakers—is yet another spur for urban economic redevelopment. But which artists—indeed, which cities—benefit from such strategies? And are there other ways to think about this relationship between the arts and place? Such questions are vital to our search for a new, more equitable approach to placemaking.

Creating Place

(*Previous spread*) Neuzz's Living Walls mural (with a self-professed urban cowboy) is part of Atlanta-based Living Walls' robust community mural program.

The installation of a new mural in downtown Detroit.

As we have seen in the efforts of Dan Gilbert in Detroit, art and culture are often at the center of mainstream creative-placemaking projects. Groups like the Project for Public Places (PPS) promote artistic undertakings because of what art—and, particularly, public art—can do. More than simply decorating something, art is understood to reinforce a locale's identity, establish the special character of a place, and facilitate community interaction. And, because new works of art are easily read as change, photographs of new installations are featured prominently on placemaking websites and in the media, likely leading to the creation of more public art. Art can play these roles and more, but the intentions of creative placemakers can at times be confused or misguided. It is not uncommon to see statements like that of a crowd-funding campaign that touted that it would be "revitalizing a community through a huge outdoor mural" or the business-improvement district that would transform "forgotten" places into "cared-for" places by installing art on chain-link fences. Regardless of these projects' other intentions, such simplified descriptions of the relationship between art and change are indicative of how little attention is paid to creative projects and their social potential.[1]

But much attention has been paid to one piece of the

economic potential of such projects. Not surprisingly, the relationships between art, culture, creative placemaking, and American cities have come to be seen as integral to urban economic growth. From Jane Jacobs to Richard Florida to the 2010 "Creative Placemaking" white paper, the "American cultural industries"—in the language offered by the paper's authors Ann Markusen and Anne Gadwa Nicodemus—have been viewed as a viable means of financial development for urban centers struggling to redefine themselves in the postindustrial era. In the past, according to Markusen and Gadwa Nicodemus: "Physical capital investments have crowded out human capital investments that hold greater promise for regional development." A greater emphasis on human capital inevitably brings one to the economic potential of the arts, "a major source of jobs." In such an equation, public art and other cultural displays become physical symbols of these relationships. Or as Markusen and Gadwa Nicodemus conclude: "Vacant auto plants, warehouses, and hotels are transformed into artist studios and housing, infusing creative and economic activity into their neighborhoods." Municipal governments have been quick to act on such ideas. Cities as diverse as Chicago; Reno, Nevada; and Carlsbad, California now have government-funded public arts offices.[2]

Not all arts-based placemakers, however, see this approach through such a rosy lens as, post–Great Recession, many artists have sought to complicate the equation that finds that art equals large-scale development. For example, there is little doubt that the financial crisis itself was on the mind of the New York–based choreographer/dancer Lydia Bell in 2009. While Novak and Flanner were establishing Eagle Street Rooftop Farm elsewhere in Brooklyn, Bell debuted the first phase of her Work for Pay project, a piece of public performance art that directly connected artists, the arts, and the world of labor. She paid a series of artists, including Sara K. Edwards, Brock Shorno, and Adriana Young, minimum wage to rehearse and perform

presentations that allowed these unemployed and underemployed individuals to show off their marketable skills, whether they be dancing, grant writing, or even crocheting. The second phase of the series took place on Labor Day in Brooklyn's East River Park as a part of the Perform Williamsburg event. The final component of the project was staged on October 3 at the Invisible Dog Gallery as part of the "No Money No Problems" exhibition, a show put on by the group Recession Art. The group's mission was to provide an "art stimulus plan" that created opportunities for artists to sell their work in new ways, particularly during moments of economic upheaval. Appropriately, the Invisible Dog Gallery had been an abandoned industrial building before it was repurposed in 2009.[3]

For Bell, Work for Pay was predominantly an artistic endeavor. However, she notes that it was also a "symbolic gesture." As a financially struggling artist herself, she meant the decision to pay the performers minimum wage to be "a statement of how we treat artists in this society." By staging the performances in public spaces, she hoped to start a conversation between audience and performer in a "very literal way." She—and others—had a strong belief that this current moment of economic dislocation could provide the means of changing "the way people think about both art and work." Work for Pay allowed both participants and observers to explore what it meant to be an artist in a "post–economic crisis" world without losing sight of the political nature of employment and unemployment.[4]

Even though the economy recovered from the Great Recession, it did so unevenly. Despite a recuperating housing industry and record corporate profits, many of the concerns raised by Bell and others persist. This situation is still fraught with peril for artists; traditional avenues of funding are uncertain, and many Americans have little income to devote to supporting the arts. Rents on lofts and other creative spaces have been increasing for years in cities like New York and Portland with the arrival of the affluent Floridian creative class, and this migration brought inflated housing costs. Galleries—outside a select group of "blue-chip" dealers—are feeling a similar financial pinch. Yet Bell sees this moment of crisis as also a moment of opportunity. The crisis illustrated that art can be created in many places, that artwork and the creative processes "don't have to happen in the studio." Out of necessity, artists must now be committed to "exploring different models of what can sustain an art project" as well as what can provide a decent standard of living. Here, perhaps, is a way to see how the arts can lead to a more inclusive brand of placemaking, one focused on the causes and consequences of social problems.[5]

Such tactics seemingly worked for Bell's performers; all three found employment following the show, with Young taking a grant-writing job offered by an audience member who was impressed by her performance. The novel avenues of sociability created by the performances allowed for the establishment of new social networks that led to economic opportunity. And, at approximately the same time, two other New York artists, Katarina Jerinic and Naomi Miller, opened The Work Office next to an IRS office in Times Square in July 2009. Using the New Deal–era Works Progress Administration (WPA) as its model, The Work Office sought to "make work" for artists by paying them $23.50—the weekly wage of an artist employed by the WPA—to document daily life in New York City. Jerinic continues to explore the intersection of public and private work, including through projects like her 2012–15 piece *Beautification This Site*. The "self-assigned residency" focused on her official adoption of a small parcel adjacent to the Brooklyn-Queens Expressway through the New York City Department of Transportation's Adopt-a-Highway program. In the project, she documented and interpreted the Sisyphean task of voluntarily maintaining the parcel on the government's behalf.[6]

These projects demonstrate that the creative aspect of creative placemaking is entangled with much more than producing art to attract and entertain the creative class. Instead, Bell, Jerinic, and Miller express the essential character of a life in the arts, with all its practical challenges of employment and production. In the process, they also demonstrate that such projects can be more than amenities to be consumed but also engines of local culture and economy that challenge the status quo. In the process, they can provide what is missing in many cities. While these artists' projects are primarily concerned with employment, other artists produce work that confronts other aspects of the cost of urban living, from skyrocketing housing prices to racial and economic segregation.

Many of these projects took form during the financial crisis, a fitting beginning for work that now must adapt to an imbalanced economic recovery *and* a new form of austerity forced not just by the declining revenues of the Great Recession but also by a federal government intent on withdrawing social support from the public sector. As local governments weigh what is and is not possible in this new era, artists have been working in the void by designing programs to improve life for those who already live in a place. Their goal is to facilitate opportunities for local residents to lead the production of something rather than attract the creative class to do it for them. Here, they follow in the footsteps of a host of do-it-yourself (DIY) sites and projects that have existed in cities for a number of decades and whose histories are often overlooked or undervalued by mainstream placemaking efforts.

But, just as making art is not simply about creativity, it is also not outside the political system. Artists must weigh the consequences of their actions, intended or not, including the well-established link between the arts and gentrification. In this chapter, we examine examples of established groups like Project Row Houses in Houston and relative newcomers like Big Car in Indianapolis that are actively wrestling with these known problems and searching for ways to identify unforeseen issues before they are unstoppable.

Making Art Work, or Working Made Art?
The Life of the Artist in Postcrisis New York City

Bell was able to stage Work for Pay through a grant from the Brooklyn branch of Funding Emerging Art with Sustainable Tactics (FEAST), which took a model much like Detroit SOUP's and brought it to cities like New York, Minneapolis, and Boston. At such meals, diners paid a sliding-scale entrance fee that granted them supper and a ballot. During the dinner, a collection of artists presented their prospective projects to the event participants. At the end of the meal, the artist whose proposal received the most votes was awarded the funds collected through the entrance fee. The winning artist was then asked to present their work at a future FEAST event.[7]

For Brooklyn FEAST cofounder Jeff Hnilicka, the decision to start an arts-based organization in 2008—it would hold its first event in 2009—"was in response to living in New York when the financial crisis hit." Understanding that "you really can't undo a global recession," Hnilicka sought a model that could replace disappearing funding sources while also critiquing those very systems of arts patronage. What emerged was an event that funded projects that often worked to beautify—or, in some cases, fix— the city in ways municipal government no longer could afford.[8]

While admitting that such an approach "was nothing new"—citing the influence of the Chicago-based Sunday Soup, a quarterly meal/arts grants event begun in 2007 by Roots and Culture, InCUBATE, and others in the Chicago arts scene—Hnilicka also found that, in 2008, "crowdfunding was not really something that existed." Kickstarter would come on the scene in 2009 (and Brooklyn FEAST shared building space with the nascent crowdfunding platform as the latter was trying to establish itself). But the key was not necessarily the funding, although that was undoubtedly useful. Instead, the goal was to empower both artist and audience. At FEAST dinners, "different people would take the reins and do their thing." This allowed artists without access to traditional funding sources to have

a literal seat at the table. And the decision to fund such work was then conducted in a democratic way, in a manner that gave dinner participants a voice in the funding process. "The audience," concludes Hnilicka, "was really impactful, as being agents."[9]

Hnilicka firmly understood that, by the early years of the twenty-first century, the concept of the creative class had come to inform the role of the arts in urban America significantly. Like Richard Florida and others, he did see a positive role for the arts in the redevelopment of the nation's cities. However, he did not see a place for authentic, grassroots arts and culture in the narrative surrounding the creative class: "We were in relation to, and a parallel to, all that." Instead, he felt a greater affinity for the DIY spaces—the punk rock clubs and underground discotheques—that had existed throughout New York for the past forty years. More importantly, the arts-based placemaking activities that went hand in hand with the creative class relied on large-scale organizations that did little to foster meaningful human interaction. For Hnilicka, FEAST was therefore "a way to do the things we were drawn to in a way that felt more human scale." To formalize such things would kill the vibrancy. As Hnilicka notes: "I was never intent on doing a 501c(3).... Not everything should be institutionalized." In fact, not everything should even be seen as permanent: "There is value in a thing emerging and disappearing."[10]

That balance between temporary and permanent is clear in the projects FEAST funded, its emphasis on actually producing things, and its long-term trajectories. The FEAST-funded artist Melanie Jelacic's 2010 "Shower Room," which involved tiling the showers at the Metropolitan Pool in Brooklyn, saw art as a means of improving an urban public space that government—or the private sector—could no longer maintain on its own. But this was no fly-by-night installation; Jelacic received official approval for the project and employed workers participating in the New York City Department of Parks and Recreation Parks Opportunity Program, a transitional employment-training program that targets sections of the city's population hit hard by unemployment. The city, as Jelacic notes, had painted the showers from time to time. "However," she continues, "no sooner than the paint has dried, mold begins to grow. It is a wasteful cycle that the city cannot afford to remedy by itself. This presents an opportunity to create a decorative work of functional art." In effect, Jelacic and FEAST Brooklyn provided the majority of the resources for a capital-improvement project. The project is still in place and presumably saving the Parks Department time and money nearly a decade later.[11]

Another FEAST grant recipient, Dan Funderburgh's "Homefront: Habitat Decoration Initiative," was more temporary. This project allowed Funderburgh to use homemade wallpaper to decorate abandoned and neglected flat surfaces throughout North Brooklyn. Unlike so many private street art projects, its intention was not to prime the space for development but to "restore a sense of ownership and habitat" to the neighborhood, whose identity was increasingly defined by absentee landlords. Like Bell, Funderburgh saw the economic crisis as allowing for a "more direct connection" between the public and artists. And the financial support offered by the nontraditional FEAST was indispensable: the project, Funderburgh noted, was "not something [he] would normally be able to afford." Such financial assistance allowed him to do something on a larger scale, a project that not only was artistically satisfying but could also "improve the general quality of life" within his neighborhood. The landscape of urban vacancy provided a new sort of canvas, and he was able to do something that others (property owners and the city government, among others) could no longer afford to do.[12]

Other artist groups further explored this relationship between economic crisis and space that was at least temporarily available. The Trinity Project, which was awarded $600 by FEAST Brooklyn in April 2010, oversaw an art and community exchange program in East Williamsburg that provided free church space for visual and performing artists in exchange for volunteer teaching at the adjacent Catholic Academy or general volunteering with Most

Holy Trinity Church. Operating with a mission statement of "connecting community through creative exchange," the project offered artists "a chance to give back to the community and build new audiences for their work, while supplying them with a crucial commodity: space": "In so doing, we begin to bridge the gap between community members who have lived in the neighborhood for generations and the artists who now call East Williamsburg home. This allows us to bring together the long standing residents of the neighborhood with the new arrivals, forging new relationships and connections in order to build a more cohesive and vibrant community."[13] Importantly, as articulated by the project cofounder Monica Salazar Olmsted, the project also offered a valuable service—arts education programming—to children at Catholic Academy from prekindergarten through the eighth grade "who otherwise had no arts education . . . if you can believe it." Here, artists worked as educators in exchange for valuable studio space.[14]

The Trinity Project was initially a success, so much so that Salazar Olmsted and her cofounder Megan Tefft wanted to replicate the approach at other churches struggling with issues of vacancy. "In order to do this," explains Salazar Olmsted, "it was important to us to formalize our operations and to be able to go for funding, which necessitated something more than a handshake agreement with the generous Friars of the Most Holy Trinity Church. We needed the security of a lease." Yet the Catholic Diocese of Brooklyn was unwilling to engage in such conversations. The church leadership that supported the project was relocated, and the school itself closed in 2013.[15]

Other religious institutions in New York City (and elsewhere) also worked out similar, temporary arrangements as they attempted to maintain large properties with declining donations caused by shrinking congregations and, often, pocketbooks. In another example, in September 2010, the "One and Three Quarters of an Inch" exhibition was staged at the former convent at St. Cecilia's Parish in Greenpoint, Brooklyn. As churchgoers in the neighborhood moved away—and died—St. Cecilia's allowed artists

Jennifer Gustavson's installation for "One and Three Quarters of an Inch."

to stage film shoots, screenings, dance performances, and art exhibitions as a means of making the parish relevant, both financially and culturally, in the twenty-first century. With the "One and Three Quarters of an Inch" show, the nunnery itself, described by one reviewer as "seductively decrepit," helped create an atmosphere that spoke directly to the show's themes of ritual, desire, mortality, and identity. Those wishing to see all the works in the exhibition had to explore more than thirty rooms over the convent's darkened four floors with just a flashlight as a guide. The exhibition's title—a reference to the size of the gap left in a window within the convent that would not shut all the way—highlighted the desire of the curator, Peter Clough, to use "such a discrete and diminutive measurement as the title for a project that had a rather epic scale." Shows like this continued for some time, but, as the economy recovered, the church signed a long-term lease with a real estate developer. The building and adjacent school have now been transformed into high-end apartments, where studios can cost more than $3,000 a month. In cities like New York, art is not always a catalyst for equitable development; sometimes it is just a placeholder until capital feels comfortable returning.[16]

But that does not mean that art cannot have a sustained effect outside these complicated urban dynamics. FEAST Brooklyn, which never saw itself as a permanent organization, held its last event in 2012, but artists funded through the program would keep alive its spirit. Dan Funderburgh, for example, has become a sought-after artist, illustrator, and wallpaper designer, creating commissioned work for such clients as the Montclair Art Museum and Le Méridien hotels. He provided artwork for Ta-Nehisi Coates's *We Were Eight Years in Power* (2017). And, in 2016, he completed a work commissioned by the Metropolitan Transit Authority for the Metro-North Railroad's Fordham Station in the Bronx. "Eureka" is a series of four windows, composed of two layers of water-jet-cut aluminum in gold and black, that greet passengers on the station's northbound platform. Following the demise of Brooklyn FEAST,

Hnilicka founded a queer collaborative art space before taking a job with Minnesota Public Radio.[17]

The legacy of groups like Sunday Soup and FEAST live on through other programs, including Detroit SOUP and Chicago's Soup and Bread. Even so, their grants are minuscule in comparison to support offered by agencies like the National Endowment for the Arts (NEA) and the National Endowment for the Humanities (NEH). Yet, in this era of government downsizing, they may come to fill in the gaps created by cuts to such agencies. When FEAST was in its prime, a 2011 Republican Party–led effort led to a $12.5 million cut to the NEA's yearly operating budget. Such attempts to curtail funding for the arts are quickly becoming nonpartisan. The next year, President Obama's budget lowered the funding for both the NEA and the NEH by 13 percent. Those budgets have yet to recover, and the Trump administration's 2019 budget even proposed the outright elimination of the agency.[18]

Working with the Clampdown: Recession Art across the Country

It should be noted that such projects did not come to life only during and after the Great Recession. Artists, particularly in communities of color and neighborhoods hit hard by deindustrialization, had been working in such ways for quite some time prior to 2008. In Chicago, the self-taught sculptor Milton Mizenberg's Oakland Museum of Contemporary Art has, for thirty years, used a vacant South Side lot to house a collection of his work. In Hamtramck, Michigan, a city five miles from the center of Detroit, the homeowner Dmytro Szylak turned his property into "Hamtramck Disneyland," a collection of images and materials (including Elvis Presley, American flags, Christmas lights, and wooden soldiers) that is housed on the tops of two adjacent garages. Szylak began the project in 1992 and finished in 1999.[19]

In Philadelphia, the artist Isaiah Zagar and his wife created Magic Gardens after moving from Peru to the city's

Milton Mizenberg Jr.'s Oakland Museum of Contemporary Art in Chicago.

renowned South Street in 1968. Soon after arriving, Zagar had covered their entire row house with mosaics. The couple subsequently purchased and rehabilitated more buildings in the vicinity that featured large exterior walls as these spaces provided ample space for Zagar's South American–inspired folk art. In 1994, Zagar started working on what would become Magic Gardens in the vacant lot adjacent to his studio. In 2002, the Boston-based owners of this once-derelict lot attempted to sell the space. Members of the community surrounding Magic Gardens came to the assistance of Zagar and helped him incorporate the space as a nonprofit organization. The Magic

Gardens site now consists of a fully tiled indoor space and an outdoor mosaic sculpture garden that takes up half a block on a revived South Street.[20]

In Detroit, the artist Tyree Guyton's Heidelberg Project—founded in 1986—has created a multiblock artistic landscape that has shifted Guyton's status from pariah to honoree. While two instantiations of the project were demolished by the city, the project's current home is now recognized as a significant cultural site and afforded some protection by the city. According to Guyton, through the Heidelberg Project "vacant lots literally became 'lots of art' and abandoned houses became 'gigantic art sculptures.'"

(*Facing*) Dmytro Szylak's Hamtramck Disneyland.

(*Above*) Kingsessing Morris Men dance at the Philadelphia Magic Gardens.

The Heidelberg Project's "Party Animal House" and "Penny House."

The project now receives an estimated 275,000 visitors a year and illustrates that arts-based placemaking is anything but a recent phenomenon.[21]

Yet the example of Guyton is suggestive of the ways the recent financial crisis has inspired endeavors to use the spaces—both physical and intellectual—created by economic upheaval to showcase a potential for arts and culture that do more than simply attempt to attract a certain type of new resident. These projects demonstrate the geographic and artistic range of projects oriented toward addressing, in part, property relations and the use of buildings in service of community while also providing work and creative opportunities for residents.

Perhaps most practically, the Great Recession led funders who saw in such efforts something akin to a proven track record to engage further with individuals like Guyton. Funding for the Heidelberg Project began to pick up during and after the financial crisis as it received substantial grants from the Kresge Foundation (2009), the Community Foundation for Southeastern Michigan (2010), the Rauschenberg Foundation (2012), Bloomberg Philanthropies (2015), the Michigan Council for Arts and Cultural Affairs (2016), and the Erb Family Foundation (2018). At the same time, the economic crisis allowed Guyton to reposition the group's impact on Detroit. The project became more than a means of beautifying a decaying neighborhood; it also came to be a point of artistic production itself as well as an educational and job-creation endeavor. By 2011, the Heidelberg Project was estimated to have a local annual economic impact of $2.8 million while employing over forty people. In early 2018, it initiated the Heidelberg Arts Leadership Academy, a free in-school or after-school arts education program meant to introduce students near Heidelberg to the potential of becoming artists and other sorts of "active change agents in their community." Later that year, the project announced its plans to open a new headquarters to house such activity along with a café, a gallery, and an event space.[22] Establishing such a space on the city's East Side fore-

grounds Guyton's art as a catalyst for yet another kind of grassroots-based development.

Others across the country have followed similar trajectories. In Cleveland, the artist Ivana Medukic's Project Pop-Up Galleries started by transforming the interiors of abandoned residential and commercial properties in the Collinwood neighborhood into temporary art galleries. Rather than simply being storefronts for viewing art, much of the emphasis was on getting visitors into the buildings to interact with the spaces. One such gallery was staged in a former dentist's office (complete with an old examining chair and X-ray machine) owned by Northeast Shores Development Corporation, a nonprofit community-development organization. In 2012, the two groups partnered with other neighborhood arts organizations and won a $500,000 ArtPlace grant to formalize their efforts as Collinwood Rising, reinforcing communitywide partnerships in the process. They followed the 2012 grant with another in 2015 designed around democratizing arts investments by formalizing community input throughout the design, investment, and implementation process. Art would be, not installed for the neighborhood, but created by it. And, while the Cleveland case is exceptional, pop-up galleries of varying scales and purposes continue across the country, from Seattle to Birmingham, Alabama. If nothing else, such pop-up galleries illustrate how the financial crisis created a space where the relationship between property and art could be reconsidered.[23]

It is obvious that money from ArtPlace America comes with certain conditions, and many projects do not necessarily fit in with its approach to placemaking. Some have chosen to avoid such funding, with varying degrees of success. In Baltimore, the Copy Cat Building has served as a sanctuary for the city's underground artists for close to thirty years. An industrial warehouse (it was once occupied by the Crown Cork and Seal Company) turned residential studio space, the building housed a variety of light-industrial tenants when purchased by Charles Lankford in 1983. As these tenants followed the broader industrial exodus out of Baltimore, Lankford began to experiment with the space by converting one floor into artist studios. Soon, many tenants, including musicians, photographers, and artists from the nearby Maryland Institute College of Art, had turned the building into a site of production as well as a living space. Over time, it came to house many of Baltimore's cultural producers: the renowned Baltimore musician Dan Deacon and his Wham City Collective once called the Copy Cat home (there is even a rumor that Deacon buried his dead cat in the building's ceiling). Cavernous living/working spaces within the structure have been rebuilt countless times, often with found materials. For many years, in an effort to avoid run-ins with city officials who did not know what to make of the space, Lankford attempted to keep the project off the radar of local politicians and policymakers. But he quickly realized that such a model was neither safe nor sustainable. He has always strived for code compliance, and, in 2003, he applied for a planned-unit-development ordinance. In 2017, Copy Cat tenants pushed the city to revisit its zoning regulations. Such pressure led Baltimore to establish a new industrial/mixed-use designation, allowing for live-work spaces in formerly industrial buildings.[24]

Similar models that attempt to remain off the grid have not been as successful. The most tragic example is the Ghost Ship disaster, in which thirty-six people were killed in a fire in a live-work artist space in a converted warehouse building in Oakland, California, on December 2, 2016. Even though the building's tenants had made significant alterations to the building, building-code-enforcement inspectors had not been inside the structure for thirty years. Owing to a lethal combination of silence on the part of tenants (they did not want to draw unnecessary attention to their activities) and diminishing city funding for actual building inspectors and inspections (an investigation by the *Mercury News* showed that 80 percent of unsafe-building notices issued by the Oakland Fire Department went unchecked by code inspectors), the Ghost Ship has become an all-too-real warning for DIY artist-based working and living spaces.[25]

The Ghost Ship tragedy produced a ripple effect among other spaces as similar sites closed down in the immediate aftermath of the fire. In San Diego, for example, the Glaushaus Artist Collective Warehouse, located in a previously vacant building in the city's Barrio Logan neighborhood, was forced to close in August 2017 because of safety and fire concerns and code-compliance violations. The space, which opened in 2009, had grown to include twenty-one studios, a public gallery, and an event space. None of the interior construction creating such amenities was, however, legal.[26]

Yet, though also taking advantage of opportunities presented by financial upheaval, other, similar sites have taken different, more formalized approaches to the creation of such spaces. In Providence, Rhode Island, AS220, a nonprofit community arts center founded in 1985, has worked to make the city's downtown neighborhood a hub for artistic performance, education, and production. Starting as a one-room artist space catering primarily to countercultural artists and underground punk acts, it was transformed in the early 1990s from a derelict twenty-one-thousand-square-foot building into a multifaceted headquarters. AS220 has continued to grow in the twenty-first-century century, as seen through its purchase in 2008 of the Mercantile Block, a well-known historic building in Providence, as the bottom fell out of the local real estate market. Within these buildings, the group books all-ages music shows, stages art exhibitions, provides studio space, and runs a café.[27]

Looking at the group's history, it is interesting to note that its moments of expansion have often followed moments of economic crisis. In 1992, in the aftermath of the recession that gripped the United States during the early 1990s, AS220 acquired an abandoned property at 95-121 Empire Street that, following the turn-of-the-century dot-com bust, it renovated. And the recession provided another avenue for the group to grow once again, as seen in its 2008 purchase of the Mercantile Block property. Such a development strategy has allowed AS220 to do more with less, but it still requires funding. To meet

Greg Brotherton works on a new sculpture in Glashaus.

such a need, the group has sought and received fund-ing from the Ford Foundation, the Kresge Foundation, the Surdna Foundation, and the Rhode Island Council for the Humanities, among others. But it has been able to use this funding to strengthen its commitment to underground arts and culture and to expand programming to address other marginalized and underserved populations. It has, for example, pioneered an arts-training program for re-cently incarcerated young people and continues to put on shows by bands with names like Freakbag, Twin Drugs, and Frenzy of Tongs.[28]

It is with such work in mind that AS220 has come to position itself as an alternative to both the informal strat-egies that inform projects like the Ghost Ship and more mainstream understandings of creative placemaking as offered by the likes of Richard Florida and PPS. In fact, it has even come up with a name for such an approach. As part of its thirtieth-anniversary celebration in 2015, the group hosted a panel on the topic of "authentic place-making" featuring AS220 founder Umberto Crenca, the Council for New Urbanism's Andrés Duany, and Project Row Houses' Rick Lowe—whose work will be discussed later in this chapter—among others. For Crenca, place-making must be democratic. What this means for artists is that placemaking must be committed to providing an arena for uncensored, unjuried work. Yet democracy in placemaking also means that, in terms of both finances and tastes, such places should be open and accessible to everyone. According to Crecna: "[People] want to go to cultural events that cost $5, and they want to go to cultural events that cost $100. They want experimental and funky and underground, and they want traditional and predictable." For the alternative space that AS220 strives to be, this means providing programming that more mainstream organizations often shy away from. Importantly, such a philosophy provides a vital counterhistory much like the one we documented for urban farming in chapter 4, one that allows practitioners to see another way to approach arts-based programming and placemaking.[29]

From Visual to Participatory: Murals, Abandonment, and the Politics of Public Art

Yet it is probably murals that have come to be the primary examples of public art in American cities. As alluded to in the beginning of the chapter, these projects are usually a safe form of arts-based placemaking, one that serves as a sort of billboard for the creative class. Often sponsored by companies courting young customers, they range from murals that celebrate communities with some kind of quirky or ironic nod to advertising cloaked as visually compelling installations. For example, in Milwaukee's gentrifying Bay View neighborhood Pabst Blue Ribbon has installed murals, and in Minneapolis Bon Iver installed a mural on the side of a skateboard shop that would become the site of a public listening party for the band's 2016 album *22, a Million*.[30]

Well before Pabst was designing ads to appeal to hipsters, murals helped define the physical and visual landscapes of places. For example, in San Diego's Barrio Logan neighborhood, murals went hand in hand with the development of the community's Chicano Park. This park space had been occupied by members of the Latinx community in 1970 after city officials backtracked on a plan to provide the land to the neighborhood for a park. After

much struggle, the park became a reality in the spring of 1971.[31]

Murals soon became a way to mark the importance of Chicano Park to the community, to document this often-overlooked history, and to assert both community identity and ownership of the space. Building on the traditional role of Latin American public murals, a crucial component of such campaigns was the Chicano Park Monumental Public Mural Program, conceived by the community artist Salvador Torres in 1969. After the community-led takeover of the space in 1970, the mural program became a reality in 1973 with the completion of a mural featuring Quetzalcóatl, the mythical feathered serpent of pre-Columbian Mexico. Later murals featured such individuals as Don Pedro Gonzalez, the owner of the first Spanish radio station in California.[32]

Such murals came to fill gaps located in communities like Barrio Logan, providing opportunities for expression along with the reclamation of both history and valuable urban space: the Chicano Park murals exist in a space that was once meant to become a parking lot. As Arban Quevedo, an artist who participated in the Chicano Park mural efforts, explained: "What happened before when officials came in and destroyed a barrio is that people took their hats off and went to a new barrio. But what I say is 'What did I go to school for? Damn it.' We have a lot invested here. I don't want to take my hat off. I want to leave it on my head and rejuvenate the barrio." Murals thus become a way to stake claims on both historical and physical territory, bringing to the surface narratives often overlooked or destroyed as development moves into a community.[33]

Across the country, murals walk a fine line between public and private forms of artistic expression as the artist's individual intent must coexist with community-based interpretations of the mural's content. And, as seen above, such pieces may also be marked by a blurring of the boundary between the public and the commercial use of space. These roles are enhanced in communities con-

Mural and listening party for Bon Iver's *22, a Million*.

tested by demographic and cultural transitions, where territory is measured block by block. Still, murals can have a double edge. For potential gentrifiers in Barrio Logan and the Chicago Mexican neighborhood of Pilsen, the murals' seemingly benign presentation to outsiders can also work to provide a shallow authentic feel to a place slowly losing all but the public elements of a displaced community.

Yet, as the Chicano Park example illustrates, such murals can be more than just evidence of a vibrant arts scene. Because of the historical significance of the pieces, the park was designated an official historic site by the San Diego Historical Site Board in 1980. The park itself, located in a city where real estate has exploded in value since the 1970s, was listed on the National Register of Historic Places in 2013. Three years later, it was designated a National Historic Landmark. Such designations keep the murals up and the space public even as the neighborhood is rapidly changing. Both these outcomes allow community members to continue to connect with the histories on display in these murals while they use the park for a variety of neighborhood-based events. Such local performers as Ballet Folklórico CalifAztlán, Karina Frost and the Banduvloons, the Quetzalcoatl Band, and DJ Viejo Lowbo

Playing basketball in Chicano Park.

The *Dr. J* mural, by Kent Twitchell.

have all performed in the park, often during the annual Chicano Park Day festival. At this event, myriad arts and crafts and food vendors peddle their wares. In real ways, the murals now anchor a space that provides avenues for indigenous cultural expression, economic opportunities, and the delivery of services for community residents. This more robust approach to murals is a good way to think about their role in a more inclusive understanding of creative placemaking.[34]

One sees how this can play out in similar ways in other cities across the country where local governments have played key roles in constituting public arts endeavors. The largest of these public arts programs is the City of Philadelphia Mural Arts Program. Established in 1984, the program began as a piece of the city's Anti-Graffiti Network and employed the muralist Jane Golden as a liaison of sorts between Philadelphia government officials and the city's graffiti artists. The tumult of the 1960s and 1970s had left Philadelphia with a seemingly endless supply of surfaces to paint. Not only did the city have the large-scale walls found in any metropolis along train routes and in industrial buildings, but its preponderance of row houses meant that, when a building was demolished, two more blank walls were created. The program's goal was to direct writers' efforts away from the illegal tagging of these buildings and toward mural painting in a straightforward attempt to beautify and improve neighborhoods suffering from blight.[35]

Owing to the success of the program, Mayor Ed Rendell announced in 1996 that the Anti-Graffiti Network would be reorganized into the Mural Arts Program, with Golden at the helm. Golden also established the Philadelphia Mural Arts Advocates, a nonprofit organization designed to raise funds for the fledgling Mural Arts Program and work with the city in an attempt to create a working private/public partnership. Since its inception, the Mural Arts Program has created over three thousand murals and, in the process, shifted from a beautification program to something much fuller. The program now annually serves fifteen hundred young people at neighborhood sites across the city. At the same time, it reaches out to prisons and rehabilitation centers, allowing inmates and others to receive a stipend to create murals for schools and community centers while creating both new social networks and economic opportunities for this vulnerable demographic. Philadelphia Mural Arts also offers a reentry program, called the Guild, for those recently released from prison. This paid apprenticeship program trains participants in scaffold installation, mosaic tiling, landscaping, carpentry, and mural and wall repair and installation.[36]

As such programming suggests, Philadelphia Mural Arts produces works of art that go beyond decorating unpainted walls. One sees further evidence of the multifaceted nature of these works in the ambitious case of "Healing Walls." This project was originally conceptualized in 2004 as a single mural to be designed as a form of collaborative restorative justice with survivors of crime and current prison inmates. Conflicts between the groups split the mural into two compositions, but the two groups collaborated on the painting process for each design. Devised by Cesar Viveros and Parris Stancell, the walls now stand on the same block, addressing the neighborhood as a compound sentence. Stemming from such work, Philadelphia Mural Arts began the Restorative Practices youth program to provide arts education to students in foster care and in the juvenile justice system. Here, arts-based exercises are combined with trauma-informed care and restorative practices as a means of helping young people cope with difficult circumstances. As state-sponsored providers of similar services face shrinking budgets, this work provides ways such populations can address the very real repercussions of trauma in urban America. The ability to address place-based trauma directly—something that more mainstream placemaking endeavors do not offer—suggests just how powerful a more grassroots approach to the practice can be.[37]

Even when Philadelphia Mural Arts murals are the product of a more typical design and painting process,

Healing Walls (Victim's Journey) and *Healing Walls (Inmate's Journey),*
by Cesar Viveros and Parris Stancell.

they often address the very issues swept under the rug by typical placemaking initiatives. Rather than producing designs that might be appealing to the creative class, they commemorate important civil rights actions and confront violence. And, while projects may celebrate neighborhoods, they often do so in a way that connects the past to the present rather than holding up the past to add character to a place without an identity other than new or luxurious.

Across the country, the impact of the financial crisis on cities has only increased the appeal and sheer number of murals, even years after the Great Recession was deemed over. Yet the economic climate has simultaneously made it harder for city governments to fund such programs as the Mural Arts Program. To fill this vacuum, grassroots organizations have taken it on themselves both to fund and to create such works of art. In Oakland, the Community Rejuvenation Project (CRP), a collective of street artists, has risen to prominence through an almost militant commitment to bringing public art to the city through murals. According to the project founder, Desi W.O.M.E.: "The idea of rejuvenation is that we're targeting the most blighted areas; the places that the owners aren't taking care of. We're going out transforming these places to give them new life." While founded in 2005, W.O.M.E. and other artists began targeting such structures throughout Oakland throughout 2008 and signing their collective work with "Community Rejuvenation." In 2009, the group hosted a summer youth employment program through funds provided by the Lao Family Community Development group, a nonprofit organization that serves low-income neighborhoods throughout Oakland. With this funding, CRP was able to hire thirty Oakland youths to clean up 150 blocks in the areas where the group was going to begin painting murals.[38]

Since then, the group has painted hundreds of murals throughout the East Bay. Some featured themes like overreliance on fossil fuels and police brutality, others such historical figures as Marcus Garvey and Rosa Parks. At Twenty-Fifth Street and Martin Luther King Way, an entire vacant building was painted with a series of images, including indigenous people playing drums, a child on a tricycle, and a snake holding a sign that says "Stop Driving." Yet, in many ways, the process of creating these murals is more important than the content; it is about taking control of one's surroundings and a belief that painting on an abandoned building can turn it into something else. The creation of such murals can also be a way to assert alternative understandings of use of space—and even ownership of such spaces. Or as W.O.M.E. explains such matters: "We feel that the community, in the same way that you can adopt a median, should be to adopt the wall. Instead of continuously making the city responsible, which costs taxpayer money, the community should have the option of saying, 'We're going to maintain this wall.'"[39]

One of the group's significant earlier projects, the Funktown Arts District, brought four new murals to East Oakland in rapid succession. Neighbors noted that these projects fill a void created by the financial crisis. "The murals brought color to the neighborhood after the Parkway Theater closed down," comments Mike Melero, a barista at Woody's Café, a coffee shop across the street from the abandoned theater. CRP has continued to build on that idea, from its "Decolonize" mural, which uses traditional Mayan and Aztec images and symbols to call for a return to a simpler way of life, to murals dedicated to Oakland's tricked-out scraper bike culture along the East Bay Greenway. The Greenway's high-profile transformation of the underutilized areas along a Bay Area Rapid Transit line is an example of CPR's "pavement-to-policy" orientation, meaning that the group not only works in the streets but also attempts to reshape Oakland's political establishment through advocacy work. The wall draws attention to the cultural history of the community while also being another proving ground for the idea that street art can bring the community together rather than simply being another space fought over by taggers and antigraffiti enforcers. It is a brightly colored symbol that the neighborhood is ready to be consumed. CRP's hope is that a street art–positive policy will keep young artists out of jail and

Funktown Arts District mural.

help reinforce community heritage and identity, particularly in a city where gentrification is rampant.[40]

As the Funktown Arts District was getting off the ground in Oakland, in August 2010 the city of Atlanta hosted "Living Walls, the City Speaks," a conference on street art and urbanism that was meant to provoke discussion of the problems facing the twenty-first-century city and how public art could possibly help address such issues. Thirty-six artists, including muralists, graffiti writers, and others, came to the city to rethink how public space should look and work ideally. They began conversations on both informal and formal ways to deal with city governments and how to get the public on the side of public art.[41]

One artist who was an early participant in Living Walls was Hugh Leeman, a San Francisco–based artist who had been slated to put up a mural of Martin Luther King Jr. on the wall of the Sound Table, an Atlanta restaurant adjacent to the Martin Luther King Jr. National Historical Park. The owner of the building that housed the restaurant had sold wall space to an advertising agency representing Old English malt liquor, but the agency would not allow any mural, even one of Martin Luther King Jr., to go up over the malt liquor ad. This reality led Leeman and Sound Table to downsize the mural, putting it up next to the ad along with the notification that "the Sound Table and muralist Huge Leeman neither endorse, condone, nor benefit from

Hugh Leeman's Dr. Martin Luther King, Jr. Mural for Living Walls, without ad.

this advertisement." Soon after the mural went up—and news of the controversy broke—the advertising agency painted over the offending ad. On the one hand, this episode demonstrates how a very specific understanding of property rights still informs much of the built environment of the American city, even during this moment of crisis (and how the mechanisms and institutions that have come to undergird such understandings remain in place). Yet one cannot overlook that, in this specific case, community-sponsored art won out over the type of commodity exchange that had previously marked this urban space. It is a success that Living Walls has attempted to duplicate—without the initial conflict—ever since.[42]

Yet, during the recent crisis public art, making artistic statements on structures has become more participatory and political and has even been employed by city governments themselves. In Milwaukee, for example, the foreclosed homes surrounding the abandoned A. O. Smith/Tower Automotive site, a multiblock complex on the city's North Side, have presented an opportunity for the city itself to repurpose such structures. For much of the twentieth century, the auto frame–making division of A. O. Smith provided employment for many members of Milwaukee's African American community. Additionally, the firm's eighty-four-acre complex, located within the North Thirtieth Street Industrial Corridor, served as a spatial reminder of the importance of manufacturing to the city's economy. Like many of its competitors, the company suffered as deindustrialization hit Milwaukee, and, in 1997, Minnesota-based Tower Automotive bought the facility. Yet this change in ownership did not stem the tide of job loss: by 2003, employment at the plant had shrunk by more than 60 percent. The last truck frame produced on site left the facility in 2006. Today, the city of Milwaukee owns much of the site—renamed Century City—and is attempting to remake it for the twenty-first century.[43]

What is most interesting about the effort to sell the Century City site is the way the city has used art as a means of making the space more attractive. In 2010, the

Thirtieth Street Industrial Corridor Corporation and Business Improvement District Number 37 invited IN:SITE, a Milwaukee-based organization dedicated to bringing public art to the city, to install art projects in the region surrounding the former A. O. Smith/Tower Automotive site. Here, public art is meant both to beautify a derelict facility and to attract investor attention to the redevelopment effort.[44]

In Chicago, the artist Margarita I. Alvarez's 2011 "Boarded" project used public art staged in vacant homes in the city's West Side Humboldt Park neighborhood as a means of creating a template for future abandoned-space interventions, including one in San Juan, Puerto

The A. O. Smith installation.

The Crown Royal Bag house
from Amanda William's
"Color(ed) Theory" series.

Two of Christopher Green's murals for Better Family Life.

Rico. Elsewhere in Chicago, the artist Amanda Williams has been producing work at the intersection of African American culture, commerce, and the built environment that draws from her experience growing up on the South Side. Her celebrated "Color(ed) Theory" series from 2014–16 transformed derelict houses with vibrant installations evoking colors of commercial products targeting African Americans, highlighting how racist development policies produced landscapes of abandonment in the city's Englewood neighborhood. The project was eventually featured in the 2015 Chicago Architecture Biennial and has been the foundation for other projects, including Williams's 2017 collaboration with Andres L. Hernandez. That project, which was mounted in St. Louis, activated and explored the history of a building that was slated for demolition.

Not ten blocks away from Williams's and Hernandez's St. Louis project, the not-for-profit organization Better Family Life has since 2016 been working with the local artist Christopher Green to install murals in the open windows and doors of derelict houses along Page Avenue. Green, who lives on the street, paints prominent Black St. Louisians—from Maya Angelou to the boxers Leon and Michael Spinks—to draw attention to people who gave back to the city. While Green and Better Family Life see the murals as part beautification project, they are also an intentional complement to the organization's other efforts in the neighborhood, including violence deescalation programs and antiforeclosure work. To that end, Better Family Life used the unveiling of the mural project to encourage neighbors and other St. Louis residents to maintain the derelict properties through cleanup days and other organized activities.[45]

Production from Dereliction: Art, Jobs, and Material in the Remains of the Twentieth-Century City

In Buffalo, the artist/architect/educator Dennis Maher takes this interaction with the landscape of abandonment further. His approach started developing through such on-going projects as Undone-Redone City, in which he used the fragments of abandoned structures to create a final product that can best be described as part sculpture, part visual history, and part political statement. Starting from the belief that "demolition is a form of cultural erasure," he literally builds on the "textual potential" and "suggestive possibilities" of "matter that was discarded" in his "work with the literal fabric of the city." There is a powerful artistic component to the work—a cube of material that once made up a home can be awe inspiring. Yet Maher's work also gets observers thinking about adaptive reuse and finding new ways to think about how a city should ultimately look and work, all the while seeing novel uses for what many would simply consider waste: "[A] rethinking of [materials that literally make the urban built environment] leads to a rethinking of [how this urban built environment should function]." Maher also sees his art as a means of addressing architects and designers, to motivate them to reengage with American cities. By using materials salvaged from actual buildings in Buffalo, he is urging his fellow professionals to "move that form of paper architecture into the natural fabric of the city."[44]

Over the last several years, Maher's commitment to this reconstitutive work has grown into an artistic and organizational practice. His more comprehensive approach is still rooted in the built environment but now also directly confronts Buffalo's inequities. The first such effort was the Fargo House, one of two West Side Buffalo buildings he saved from demolition in 2009 and transformed into sites for architectural and curatorial experimentation. While the smaller of the two buildings is still in flux, the larger currently houses his residence, two rooms for visiting artists, a gallery, and his extensive collection of objects he acquires through demolitions, flea markets, and donations.[45]

The second is Assembly House 150, a not-for-profit organization that operates out of the former Immaculate Conception Church, just over a mile from the Fargo House. Although the building's congregation had been out of the church for only a few years, the place was derelict and without heat. At first, the church was an experimental

The second floor of Dennis Maher's Fargo House.

(*Above and following spread*) Dennis Maher sets up Assembly House 150 in 2014, 2016, and 2019.

Dennis Maher works with the Society for the Advancement of Construction-Related Arts team on a new porch and entrance for an affordable housing unit.

extension of the Fargo House, Maher's workshop, and a space where University at Buffalo students could collaborate. It has since grown to be the home of the Society for the Advancement of Construction-Related Arts (SACRA) as well. The initiative is a collaboration with the Albright-Knox Innovation Lab and funded in part by a $100,000 NEA Our Town grant.[48]

The project is an outgrowth of a project Maher initiated while he was the 2012–13 Albright-Knox Gallery artist in residence. For his installation "House of Collective Repair," he invited eight tradespeople (plumbers, masons, and electricians) from the Buffalo area to make model houses out of the materials they use in their work. These sculptures were then integrated into Maher's overall installation. Extending this experience, SACRA has two main goals: (a) "Create teaching/learning opportunities where exceptional quality and richness of imagination in design and construction are encouraged." (b) "Teach the necessary skills for employment in the fields of carpentry and woodworking."[49] Rather than working with employed tradespeople, the program works with candidates identified by the Erie County Department of Social Services in a novel skill-development program that has participants constructing sculptures inside the church and working on commissioned projects. Recent commissions have included a wall-sized art installation for the Northland Workforce Training Center, a major new job-training center on Buffalo's East Side, and an affordable-housing rehabilitation project for PUSH Buffalo, a housing-advocacy group and not-for-profit developer. In so doing, SACRA directly addresses an unemployment rate that is twice the nation's average while also working to protect the city's built environment.

Across the country, artists and organizations in other cities are making use of the remnants of abandonment in similar ways. Since 1994, the Green Project in New Orleans has sold scrap from, in its language, "decon-structed buildings." And its mission statement—"reduce, reuse, reclaim, repair, recycle, repurpose, restore, rehabilitate"—could have become the mantra for postrecession American cities. Through the Green Project, such things as lumber, bricks, nails, paint, and other materials have been used to renew homes throughout New Orleans. But there is also an artistic component to the group's mission. It has sponsored events like the Salvations Salvaged Furniture Design Competition, in which local artists and designers use materials purchased from the Green Project to create furniture.[50]

In San Francisco, Scroungers' Center for Reusable Art Parts (SCRAP) is even further committed to the idea

A Green Project worker removes nails from donated wood.

of recycled materials being used for the arts. Since 1976 (when, as the its website makes clear, the group came together "to provide art supplies to art teachers in the San Francisco public school system during America's last serious financial crisis"), SCRAP has been offering such materials as paper, plastics, woods, buttons, textiles, and craft and office supplies for what the group terms *creative reuse*. These supplies are then made available to teachers, parents, artists, and various organizations at cut-rate prices. During the economic crisis, SCRAP took to running workshops on such topics as mosaics, sculpture, and screen printing and continues the practice today.[51]

While such work is rooted in the arts, it can also lead to spaces of production where new narratives emerge. In the process, the reductive creative-class approach can be challenged as identities, ideas, histories, and tensions often ignored by mainstream creative-placemaking endeavors—including those associated with race, underground culture, economic inequality, and gentrification—are brought to the surface. This allows individuals and communities often left out of the mainstream creative-placemaking discourse to join the discussion in a way that brings much-needed attention and resources to the spaces that need them the most. In that process, one sees a more democratic, more inclusive brand of placemaking beginning to take shape.

"I Wish This Was . . .": Postcrisis Public Art, Urban Planning, and Civic Participation in the Twenty-First-Century American City

Such artists as Amanda Williams and Dennis Maher and such groups as the Green Project and SCRAP allow us to think anew about the materials that have gone into building our cities. They remind us of what has come before, of the continued vitality of our urban centers, and of the role that art can play in our lives. Yet there are other organizations and artists pushing these processes even further. For such actors, the arts—particularly at this moment of

economic dislocation—represent an opportunity both to physically renew and to plan neighborhoods in ways that look different than past efforts. One way is to suggest new uses for old buildings, with art providing the aesthetic or the activities that hold such reimaginings together. The arts can also evoke better ways to integrate citizen participation into city planning. Here, reimaginings of the urban landscape are not limited to structures. Neighborhoods—indeed, entire cities—can be rethought.

The Big Car Gallery in Indianapolis is illustrative of many of these trends. The gallery was founded as an artist collective in 2004 and quickly became an anchor tenant in the Murphy Art Center, a former furniture factory in the Fountain Square neighborhood. The gallery played a crucial role in the renewal of the neighborhood and became the first gallery in the area to open new shows on the First Friday schedule, with some exhibitions drawing close to fifteen hundred people on opening night. As it grew in popularity, it came to serve as both a commercial and an artistic hub within the neighborhood and started to plan publicly oriented art projects throughout the city. But, by late 2011, displaced by rising rents, Big Car left its space and established projects in other regions of the city, including redesigning and offering programs in a vacant automobile service station in a moribund mall parking lot. The group took the next few years to expand its geographic coverage of the city by helping neighborhood organizations realize their own projects and establishing a client-based placemaking consultancy.

In May 2016, motivated by its early experience in Fountain Square, Big Car launched its most ambitious project yet: the redevelopment of a former milk-bottling plant and metal tube manufacturer just outside Garfield Park on the city's South Side. One of the biggest lessons of its displacement from Fountain Square was that it was partially responsible for that displacement. Its years of nurturing a community of artists exacerbated the area's gentrification. Accordingly, the organization's executive director and cofounder, Jim Walker, is mindful that "people can get pushed out" when an arts organization moves into the neighborhood, so Big Car has initiated programs to limit the negative effects of its new projects.

As Big Car head curator and cofounder Shauta Marsh acknowledges: "Unfortunately, we will change the neighborhood and make it a little more expensive." But, in response to that inevitability, she notes: "We also have the housing program to make the neighborhood less expensive—long term affordable housing in an arts community."[52] Knowing that public awareness of its plans for the Tube Factory Artspace would drive up property values, before announcing the project Big Car partnered with the not-for-profit community-development corporation Riley Area Development, with support from the Indianapolis Neighborhood Housing Partnership, to purchase unoccupied buildings in the neighborhood and convert them into affordable housing. At present, five vacant and derelict houses have been purchased, renovated, and put back on the market as affordable artist housing under a land-trust-style program. Prospective residents must make no more than 80 percent of the average Marion County income, foreground public engagement as part of their artistic practice, and commit to spending sixteen hours a month doing social practice projects anywhere in the city. In exchange, they receive down-payment assistance and, in effect, purchase their houses at half price, with the Big Car–led group owning 51 percent. If they desire to sell their share in the future, the partnership will buy it back. The group has also launched a similar program for affordable rental housing with another five units.

The ten units represent a start, but the initial plan was to purchase additional buildings before the Tube Factory opened. Two years later, Big Car sought to buy Riley out of its contract. As Walker puts it: "The main thing that happened is that [Riley] needed to move on to the next thing. It's built on the past funding the future. . . . It's not by buying something and holding onto it forever. And we are committed to here. We don't want to move on."[53] The competing demands of real estate development and a long-term commitment to particular properties meant that the model was not sustainable for Big Car's not-for-profit development partner.

This organizational conflict has oriented Big Car in a new direction. Noting Indianapolis's relatively low building density as one reason for the lack of affordable housing in the area, Walker is working on a plan to "increase density in order to increase affordable housing." The group is looking at using portions of its expansive parking lot to construct mixed-use buildings with ground-floor retail and affordable housing geared toward artists above. As with the group's other affordable housing initiatives, Walker notes: "The goal will be that the housing will be filled with artists of different kinds. And, when I say *artists*, I mean barbers, cooks; I don't just mean visual, *fine* artists."[54] Residents will be required to volunteer in the neighborhood.

Of course, all this change in a low-density neighborhood has produced problems, even without the construction of new apartment buildings. So far, the biggest complaints have been about parking availability on a block that used to have fewer cars, but Big Car has addressed the issue by installing "resident parking" signs and encouraging visitors to park in its lot or take public transportation. Still, the group realizes that there is a bigger issue. As Marsh says: "A lot of people feel that contemporary art is elitist, and I've had people say to me in the past, 'I shouldn't be here, I don't have a college degree.'" In response, Big Car has "diversified its offerings." Marsh continued:

On the block, we reach out to neighbors. We give them calendar listings. We also do events just for block members so that they know that they're welcome here. It's not just a weirdo art thing. The children have always come in and play and ride the bikes in the parking lot because it's a safe place, and sometimes, if I don't have a lot of work, I get paints out, and they can paint. Their parents usually would come in [only] if people were parked in their spot, but now they're

Big Car's Spark Placemaking at work in Indianapolis's Monument Circle.

coming in at First Friday [art openings]. So, over the last two years, I think we've had a lot of success.[55]

Another thing Big Car realized about its previous move from the Fountain Square neighborhood to the middle of a mall parking lot was that those involved in the group "liked being part of a residential neighborhood and being part of the community." They're working on it. Among other projects, Big Car hosts neighborhood association meetings, commissions work that connects with the community, and operates a radio station that allows anyone to propose a show. After all, as Walker says: "You hear developers talking about being a 'developing and placemaking

company.' And all they did is make a place. Everywhere is a place. It's the making part they don't understand."[56]

The ideas of engaging citizens and reusing the built environment through the arts have also caught on in New Orleans, where the artist Candy Chang has proposed a literal road map for a reimagined city. Her "I Wish This Was" and "Before I Die" projects call for a new approach to understanding the postcrisis (and, in the case of New Orleans, post-Katrina) urban landscape. For both projects, Chang used abandoned properties throughout her neighborhood to stage her work. Startled by the vacancy rate in her community, in November 2010 she started to take inventory of what the neighborhood surrounding her

An exhibition opening and community night market at Big Car's Tube Factory Artspace.

needed (including such basic things as grocery stores). At the same time, she wanted to know what her neighbors wanted to see in their community.[57]

Casting her neighbors as makeshift urban planners, Chang designed fill-in-the-blank stickers reading "I wish this was ——" that allowed community members to re-imagine what their neighborhood could be. She then placed boxes of the stickers in businesses around the community and also posted grids of them (accompanied by a supply of permanent markers) on vacant structures, inviting passersby to write on them what they wanted to see these underutilized spaces become and then put them on the buildings. Her goal was to offer those left

out of the official discourse a new way to voice their opinions. A number of the responses were practical, including a community garden, an affordable farmers' market, and a bike rack. Others were a bit more fantastic: a number of passersby offered such responses as "Brad Pitt's house" (perhaps referencing Pitt's much celebrated—and maligned—Make It Right Foundation houses in the Lower Ninth Ward) and "a dancing school, full of nymphomaniacs with Ph.D.s."[58] The result was visually stunning, an unfiltered, flesh-and-blood representation of a community's hopes and dreams, not the prepackaged vision usually forced on people by politicians, policymakers, and urban planners.

Candy Chang's original "Before I die …" installation in New Orleans.

The next year, "Before I Die" followed a similar strategy in attempting to document the hope and dreams of those living with abandonment as their daily reality. For this project, Chang turned the side of an abandoned home into a large chalkboard on which community members could share their innermost thoughts in a very public space. The setup was quite simple: paint the wall with chalkboard paint, and stencil the phrase "Before I die I want to ——" on the side of the building. Passersby could then complete the sentence in any way they saw fit.

As in the case of Chang's "I Wish This Was," some of the answers provided were whimsical, participants noting that they wanted to "have fun" or "go 200 m.p.h." be-

fore they died. Others focused on the functional side of things: "finish school" and "raise my children" were two other responses. Finally, many used the piece of public art to offer a critical perspective on life in contemporary New Orleans. One participant wanted to "see all homeless people with homes," while another wanted to "live without money." Fitting a city beset by corruption, another wanted simply to "see the elite fall." The model has been so successful that it has become a global phenomenon, with more than four thousand installations in more than seventy countries.[59]

The work of Big Car, collectively, and Candy Chang, individually, suggests the ways placemaking can be employed to capture the ideas and voices of those usually left out of official redevelopment conversations and larger-scale urban planning efforts. Such placemaking practitioners then highlight how this process can lead to valuable interventions in underserved neighborhoods through, for example, the rehabilitation of a once-derelict structure as an art studio or the installation of a compelling, engaging piece of public art. In the process, such projects can also produce new relationships between city residents and both the creation and the consumption of art. In other words, such work can, as is additionally apparent in both the Philadelphia Mural Arts Program and Dennis Maher's work in Buffalo, provide alternative understandings of an arts-based redevelopment strategy. And the process of creating these new relationships can also provide alternative understandings of urban planning, design, use, and employment.

In the Hyde Park neighborhood of Chicago, the Opportunity Shop (Op Shop)—begun in 2009 by the artist Laura Shaeffer and her husband, Andrew Nord—attempted to rethink the urban environment in such a way. Prior to starting the Op Shop, the couple had spent two years transforming their row house into an art gallery. Running out of room, and concerned with balancing a domestic life and a public life in a single space, Shaeffer wanted to find a new place to work that also engaged the surrounding community in some real way. She began to call local building managers overseeing vacant storefronts, asking whether she could use these sites on a short-term basis for a nominal fee. She envisioned that such sites could be more than just pop-up galleries, that they could become places of community involvement and artistic exchange, all the while transforming otherwise empty urban spaces into spaces of creativity and creation. Since its inception, the Op Shop has been self-funded and managed entirely by a diverse group of volunteers who reflected the neighborhood's varied demographics.[60]

For Shaeffer, the Op Shop had to be open to the public, allowing community residents the opportunity to transform the space as they saw fit. "The Op Shop is totally public," she explained in an interview with the Chicago-based website *Gapers Block*. "I always wanted a public space. It appeals to the desire to be democratic, to be inclusive.… How do we creatively go about fostering conversation and communication in our community?"[61] In their attempt to answer this question, Shaeffer and the other Op Shop collaborators temporarily turned three other vacant storefronts into community spaces. Each iteration had a particular theme, including urban agriculture, conceptions of home, and community exchange. The vision of a broader community was developing thanks to collaborations with neighborhood partners and the explicitly intergenerational aspect of the programming. Retirees and children were core participants, along with the young artists one typically expects in such places. In the process, the group added new elements to its repertoire, including a thrift store and weekly community dinners. The energy was palpable, and the circle of regular contributors continued to grow. As Shaeffer put it: "The Op Shop happened like an eight-year-old wanting to run across a field. It was like, 'I'm going to run across this field. I'm running!' I'm not saying it wasn't hard, but it was just happening."[62] In 2011, a property owned by a local church, a former large residence turned unoccupied community center, became available near the University of Chicago campus. Shaeffer negotiated a lease that would allow the creation of a more permanent space, the Southside Hub of Production (SHOP).

Over the first few months it occupied the home, the group grew in size and formality, gaining not-for-profit status and a board of directors to help organize events and staff the space. At the same time, both temporary and more permanent engagements were staged in the space. In many ways, this new space continued the activities of the earlier Op Shops through community dinners, art exhibitions (featuring local and internationally known artists), film screenings, concerts, and the daily operation of a thrift store. The building also became a home for regular meetings of community groups as diverse as theater troupes and Zumba workout classes. Finally, SHOP featured dedicated artist studio space for rent.

Volunteers in the first Op Shop location in Chicago's Hyde Park neighborhood.

The installation of Samantha Hill's
Kinship Project: Great Migration
inside Faheem Majeed's *How
to Build a Shack* at the Southside
Hub of Production.

The band Zamin performs at the Southside Hub of Production.

Yet SHOP's longer-terms plans highlight the ways the group imagined its space as a community center for the twenty-first century, one that offered resources and services often unavailable in sections of America's urban centers and saw the arts as a vital component in both neighborhood health and neighborhood revitalization. At the time, its plan was to house a small-press library and book exchange along with a woodworking shop, a sign-up project space, a movement and bodywork space, a seed bank, and a community radio station and recording studio. For art, the space would house the Hyde Park Kunstverein—a Berlin-style community art museum featuring local collections and artists—and the CurioPantry,

an ongoing exhibition space within a Victorian pantry. For children, SHOP featured a fort-building room and a woodworking shop.

But, like so many other art pop-ups that attempt to transition into long-term locations, space remained an issue. Once SHOP's lease ended, the church shifted the terms to month to month. The crisis was over, but, as so many churches were, it was dealing with its own financial issues. Faced with such pressure, the church board was split over what to do with the mansion: continue the church's mission of supporting community engagement by renting the building to SHOP or sell the building to help shore up the congregation's finances. In the end, the financial concerns won out in a stronger housing market. Despite SHOP's efforts to secure the building, the church sold it to another buyer, and the organization had to move. The group scaled back its activities to fit inside the small storefront of what had been the city's oldest bookstore. This new incarnation took a new name, the Hyde Park Free Theater, and the group reinvented itself yet again, starting as a gallery and studio space but evolving into an eclectic community center when the donor who had pledged to support the new space for three years withdrew after financial troubles. The final straw was when the landlord doubled the rent from $1,500 a month to $3,000. Crushed, Shaeffer set up a new space, Compound Yellow, in her new home in the inner-ring suburb of Oak Park.[63] No similar organization has popped up in the neighborhood, leaving a significant lack of arts spaces that connect the general South Side and the University of Chicago community. One reason may be the double-edged success of the University of Chicago's development efforts, which have significantly reduced vacancies in Hyde Park. The time for pop-ups to transition into larger organizations may be over there, at least until the next crisis.

Founded around the same time as the Op Shop and just down the street is a now world-famous attempt to use the arts to redefine an urban community. The renowned artist Theater Gates started Dorchester Projects in Chicago's African American Grand Crossing neighborhood, creating what the *New York Times* described in 2011 as "a site of artistic and community change for a neighborhood that has suffered years of blight and cultural neglect."[64] At the beginning, the complex merged a former candy store and a single-family house into Gates's home and a hybridized cultural space that combined art with urban planning and activism.

Crucial to such a development is a fundamental rethinking of access to certain types of urban spaces. "Buildings that don't allow everybody in, buildings that don't allow for the conviviality of culture to exist," Gates has argued, "bear as much disincentive for real engagement as an abandoned building is for activation."[65] Consequently, he orients his work in the same way as the Op Shop did: toward inclusion and public engagement, a state of consistent expansion. Such an orientation situates his project, as it does many others described in this book, at a point of both alignment and tension with formalized political and cultural institutions. With the relative dearth of publicly oriented buildings—or capital—with which to create new major centers of social interaction in the predominantly African American neighborhoods on the South Side, the Dorchester Projects emerged as opportunities arose. By utilizing former residential and commercial buildings not designed for public meetings, Gates attempted to draw new life to the community in a small but also quasi-legal way. "When my alderman says 'You can't use a building like that,'" asserted Gates, "I have to say, 'There is no other building type.'"[66]

At the start, Dorchester Projects was known as much for repurposing the buildings as it was for its similarly eclectic collections: eight thousand records from Dr. Wax, a now-closed local music store that was instrumental in influencing Chicago's hip-hop scene; sixty thousand glass lantern slides from the University of Chicago's Department of Art History that were formerly used in instruction; and fourteen thousand items from the inventory of the now-closed Prairie Avenue Book Store, a once-prominent architecture-oriented bookstore in Chicago's Loop. This initial convergence of place and collection was key to Gates's mission. At the time, Gates said: "My neighborhood, like any other neighborhood that has been underresourced for a long time, has had its fits and starts of development. There have always been dedicated people in my neighborhood, folks who keep their lawns and take out their garbage and say hi to each other. It just happens to lack certain cultural amenities that would make it desirable to other people."[67]

His mission of opening up the Grand Crossing community to broader audiences through arts and culture took the form of events ranging from listening parties—at which vintage albums from the Dr. Wax collection were played while the role of music in urban communities was

The two original Dorchester
Projects buildings, the birthplace
of Theaster Gates's Rebuild
Foundation.

discussed—to public performances by such artists in residence as the jazz musician and composer David Boykin. As the programming expanded, so did Gates's vision, leading to the formation of the Rebuild Foundation, a not-for-profit organization that has pushed for further urban redevelopment in Chicago, St. Louis, Detroit, Omaha, and elsewhere.

But Chicago remains Rebuild's focus. The organization lurched from project to project as it sought a voice but has settled into a groove in the last couple of years. Its Black Cinema House is an example of the growth. In 2012, the group launched a film series showcasing and interrogating films by and about African Americans and the Black diaspora and curated by the University of Chicago professor Jacqueline Stewart. The series has been housed in three different locations, not counting the Dorchester Project's backyard. It is currently located in the organization's largest cultural project, the Stony Island Arts Bank. Opened with fanfare in October 2015 with Chicago Mayor Rahm Emanuel and other politicians in attendance, the center exhibits work by local and internationally celebrated artists among programming like that of the Black Cinema House. Special collections remain an emphasis, including the glass lantern slides collection once housed at Dorchester Projects, to which have been added the (vinyl) records of the legendary Chicago DJ Frankie Knuckles, "negrobilia" collected by Edward J. Williams, and materials from the Chicago-based Johnson Publishing Company, which publishes Ebony and Jet magazines. Its design and programming have turned the building into a cultural center as well as a tourist attraction, with regularly scheduled public tours and special events during Chicago Architecture Foundation's Open House Chicago project.

As such features suggest, Gates is not afraid to use the Stony Island Arts Bank to address issues related to race, something many mainstream creative placemakers choose to ignore. The site, for example, is now home to the gazebo where twelve-year-old Tamir Rice was shot and killed by a Cleveland police officer in January 2014. Rice's mother, Samaria Rice, partnered with Rebuild to save the gazebo from demolition in 2016. A deconstructed version of the gazebo was featured in a 2018 interior exhibition on police violence at the Arts Bank; a four-part discussion series was also initiated to foster public dialogue. In 2019, the structure was reconstructed outside the bank building. It now serves as a memorial, a historical marker, a place of healing, and a call to action all at once. To view the site is to understand why such an approach to placemaking is necessary for cities like Chicago at this moment in time.[68]

While Indianapolis's Big Car is experimenting with housing, the Rebuild Foundation has partnered with Brinshore Development, Landon Bone Baker Architects, and the Chicago Housing Authority to create Dorchester Art + Housing Collaborative. The project reuses the formerly vacant Dante Harper public-housing project to create a mixed-income community that emphasizes attracting artists to the South Side of Chicago. As Gates told the *Architect's Newspaper* in 2011: "We want to be able to attract South Side artists—artists of color—who normally have to go to the North Side to make art or make music."[69] The complex has a common arts center and thirty-two townhouses set aside for artists, traditional public-housing residents, subsidized renters, and market-rate renters.

Gates's work in Grand Crossing demonstrates how a form of cultural redevelopment can lead to the renewal of actual physical structures. What was a small, quasi-legal project founded during the financial crisis has expanded into a major organization with significant support from the political and arts elite. With it, Gates's own artistic practice has evolved from his pottery to emphasize attention to the built environment, like his 2017 installation in the Minneapolis Sculpture Garden *Black Vessel for a Saint*, which displays a statue of Saint Laurence salvaged from a Chicago church inside a brick column. Moreover, Gates

Theaster Gates with Chicago mayor Rahm Emanuel and Alderman Leslie Hairston at the opening of the Rebuild Foundation's Stony Island Arts Bank.

continues his work with the University of Chicago, where he remains integral to the school's plans to expand into the nearby Washington Park neighborhood through its Arts Incubator. With the university as a partner, Gates has been involved with everything from the arts- and architecture-focused BING bookstore to the Currency Exchange Café, whose name is taken from the Chicago term for payday-style lending outfits.

Such projects suggest how important the arts and artists can be during moments of financial crisis, particularly as cities look for innovative ways to renew the urban landscape. "A big part of my art practice has been creatively investigating what happens in neighborhoods," Gates notes. "That includes playing in the real market, not just gesturing at it. We're at a moment where the interventions that artists make are not just in museums and galleries."[70] Artists and not-for-profit organizations are simultaneously advancing such a perspective across the country, advancing grassroots efforts to engage local problems—with and without government support.

The Tamir Rice memorial at the Stony Island Arts Bank.

From Sculpture to Data: How Robocop Tried to Save Detroit (Again)

This idea of allowing individuals to have a greater stake in arts funding can also be seen in a public art project in Detroit: the drive to create and install a bust of the movie character RoboCop. In the 1987 film *RoboCop*, the title character—part man, part robot—saves a future Detroit from an outbreak of violent crime. The artist Fred Barton's bust pays homage to this fictional character while suggesting that a new sort of hero is needed to save the city from its current predicament. In less than a week in 2011, the project surpassed its $50,000 goal through the fund-raising website Kickstarter. A Facebook page helped stoke further interest (even though Detroit's mayor, Dave Bing, along with other city officials, saw little benefit in the proposed statue), and, after its forty-five-day run, the "Detroit Needs a Statue of Robocop!" campaign raised $67,436 from more than twenty-seven hundred individuals, including one donation of $25,000. As of early 2020, the statue was finally finished and set to be installed in the Michigan Science Center "in the spring or summer."[71]

Even the *RoboCop* campaigners acknowledge that it a fun and somewhat trivial project, but the language used by some of its strongest advocates, including the Detroit artist/entrepreneur Jerry Paffendorf, suggests that this

Jerome Ferretti's *Monumental Kitty*.

United States Social Forum's
temporary tent city for attendees at Spaulding Court.

piece of public art highlights a more democratic approach to artistic interventions meant to revive the landscape of the city. "I see a crowd funded *RoboCop*," Paffendorf explains, "serving as an avatar that pushes people to think differently about solving problems and creating things on a scale they currently think is impossible or [think they need] rich people or government to do for them." At the same time, there is a belief that such a project could lead to other, potentially more important interventions. "If those crazy people can raise $50,000 to create a statue of freaking RoboCop," Paffendorf continues, "then certainly and without doubt we can raise even more (more money, more talent, more willpower, more everything) to do something that we believe will make a real difference."[72]

With this approach, Paffendorf and his collaborators have initiated a range of projects that attempt to challenge how property is used and owned in Detroit. In the process, they demonstrate Detroit's creative shift in scale, from hyperlocal to regional efforts. A year before the Robocop campaign was launched, another kind of venture spearheaded by Paffendorf was making headlines. In 2010, Loveland Technologies—a digital surveying startup that tracked foreclosures and vacant lots throughout the Detroit metropolitan region—created a "micro–real estate" enterprise that sells for $1.00 per square inch a parcel of Detroit property owned by Paffendorf. The purchasers of such microproperties thus become "inchvestors." According to Paffendorf: "The inches become like little shares in the city. Even such a lightweight form of ownership has a really cool psychological effect. Even if they bought the inches on a whim, it would bring people into the city a little bit more."[73] The money raised by such property sales allowed Paffendorf to fund Jerome Ferretti's *Monumental Kitty*, a sculpture for the Corktown Pedestrian Overpass Improvement Project, support the rehabilitation of Spaulding Court, and, in June 2010, pay $1,000 for two abandoned houses (affectionately known as "Righty" and "Lefty")—one burned beyond recognition during the winter of 2009—across from Detroit's vacant

Michigan Central Station and Phil Cooley's Slows BarBQ. These two structures were rechristened Imagination Station and transformed into a nonprofit organization that hopes to turn the community into an artists' enclave.[74]

Through Imagination Station, the purchased homes became spaces for public art themselves, with projects rotating with the seasons. Lefty, the burned-out home at 2236 Fourteenth Street (the other home is at 2230 Fourteenth Street), became the artist Marianne Audrey Burrows's "Paint a Burnt House" installation in the summer of 2010. For this project, Burrows splashed a spectrum of paint throughout the damaged structure. She explained: "I put color all over it. [It was] kind of like a resurgence, like a rebirth in a way." In the fall of 2010, another Detroit-based artist, Catie Newell, created Salvaged Landscape on the site, using burned boards culled from Lefty and working while the house was coming down around her.[75]

In very real ways, such projects were meant to be short-term catalysts for both bigger and more long-term redevelopment. In October 2014, ownership of Righty—the remaining house associated with Imagination Station—was turned over to a young Detroit family. After three years of working on the house, Stephen McGee and Cory Coffey moved into what Imagination Station had termed "the Last Man Standing on Roosevelt Park," illustrating that, counter to the dominant narrative, housing could come back in cities hit hard by disinvestment.[76] Five years later, the couple is still there.

At the same time, the concept of inchvesting was meant to spur new ways of thinking about property. Loveland is now best known for producing the parcel-mapping website Landgrid, which is designed to simplify the process of collecting, managing, and displaying property information. While an online mapping company may seem to have little in common with Loveland's previous projects, the idea is a direct outgrowth of them. In the midst of Detroitmania in 2011, Loveland launched WhyDont WeOwnThis.com. The website simplified the process of browsing the more than thirteen thousand properties that

A volunteer crew renovates the Imagination Station buildings.

Catie Newell's *Salvaged Landscape* installation at Imagination Station.

Scott Hocking's installation at Gare Saint Sauveur in Lille, France.

were being sold in the Wayne County Tax Foreclosure Auction, which offers properties at least three years behind in tax payments. These properties can be purchased for as little as the amount of back taxes owed by anyone willing to bid through the county's system. If the properties do not sell in the first auction round, they are offered for as little as $500 in the second. Motivated by concerns that the properties sold in the auction were primarily going to real estate speculators with the resources to assess their quality, Loveland hoped that a website with ready access to property details could lower the barriers to entry, keeping ownership among neighbors and concerned groups, rather than absentee landlords. In the process, it continues to address local ownership and use issues, but it does so through a corporate model, with all its benefits and complications.

The artist Scott Hocking has also used the literal remnants of abandonment in Detroit to create art. His work uses salvaged materials from the city's vacant buildings to create pieces such as *Garden of the Gods*, perhaps his best-known work. *Garden of the Gods* was illegally installed on the roof of the derelict Packard automobile plant in Detroit (the renowned artist Banksy also created a work at this location). Using the structure's crumbling concrete columns as makeshift pedestals, Hocking's piece—completed in December 2009—replaced the twelve gods of the classical Greek pantheon with wooden television consoles taken from the building. Since 2009, as the building and its surroundings have fallen into further disrepair, Hocking has documented the impact of this process on his work, creating a dialogue between art and environment. This dialogue informs another conversation. While Hocking certainly shows work in traditional galleries, many of his projects connected him with nontraditional art spaces where people desire to initiate conversation about and instigate action over the issues his work raises. Among those locales is Gare Saint Sauveur, a former derelict train station turned contemporary art center in Lille, France, and Big Car's Tube Factory Artspace.

For Big Car, he based his installation on Indianapolis's long-derelict RCA television component factory, a site that has now been demolished with support from the US Environmental Protection Agency.[77]

Such pieces make a statement about the current economic climate and make the structures associated with vacancy the center of attention, crucial components of the art itself. But they also do more. They are a piece of a definable counterapproach to creative placemaking that is cognizant of similar work in Detroit and beyond.

Housing Art: Project Row Houses and the Complexity of Placemaking

Rick Lowe's Project Row Houses is perhaps the country's most famous project subsumed under the placemaking banner. In 1993, Lowe and a group of six other African American artists saved more than a block's worth of shotgun houses in Houston's historically African American Third Ward from demolition. What would develop over the next twenty-five years is an internationally recognized artist-led development that is part gallery space, part residency, part affordable-housing program, and part community experiment spanning dozens of buildings over five blocks. In the process, the organization behind the project would become an international model for socially engaged art, and Lowe would be a MacArthur Foundation grant recipient. By 2014, when Lowe received his MacArthur award, Project Row Houses had grown to include more than seventy buildings spread throughout the neighborhood.[78]

In addition to preserving homes, Project Row Houses has also started to design its own tiny homes (through the XS House program) and prefabricated affordable housing (through the InHouse OutHouse program). These structures provide housing for such underserved populations as single mothers in Houston: through the Young Mother Residential Program, single mothers between the ages of eighteen and twenty-six can received subsidized housing

Rick Lowe playing dominoes at Project Row Houses.

through Project Row Houses, providing they are looking to find employment and/or further their education. Lowe has even brought economic opportunity to the community itself. Since 2004, Project Row Houses, along with the Museum of Fine Arts Houston, has hosted the Glassell Core Fellow Artist program, which offers one- and two-year paid residencies. Leaders in such disparate places as Los Angeles, New Orleans, Dallas, and Philadelphia have brought Lowe to work on similar projects in their cities, redirecting resources toward much-needed housing.[79]

Standing inside one of the formerly derelict buildings, Lowe warns: "Sometimes you should be careful what you wish for—or what you ask for."[80] He recounts the story of the project, which begins in 1992 when a local organization petitioned political leaders to address the blight and criminal activity in the area, with the long-term goal of redeveloping the land for new single-family homes. Many of those buildings were demolished, but not before what would become Project Row Houses was established in "the worst part of the entire neighborhood," where only three buildings were occupied: two houses and a corner

Visitors stop by Project Row Houses for the opening of Round 43.

store. The idea was, according to Lowe, that "this could be the seed for an art project that would reclaim these old houses, and pay homage to the history, and bring creativity in." He continued:

And the reason I say you have to be careful what you ask for ... is that what the center was asking the city to do has put the community in a very difficult situation, right? While those houses were blight and all of that stuff was problematic, at least the structures held some semblance of what the community was, with some capacity to reinvest in it, but, instead, the clearing of the land was basically predevelopment for developers, right? 'Cause they'd much rather have land clean and cleared of buildings than to have to deal with the question of, you know, to restore something or tear it down. So the community paved the way in a big way.[81]

After years of divestment and being underserved by local authorities, the Third Ward is experiencing an unprecedented housing boom. While Project Row Houses certainly signaled the development opportunities in the area, it has also attempted to combat development. In addition to preserving the twenty-three shotgun houses, it has also partnered with the Rice University School of Architecture to construct even more affordable rental housing. About the development, Lowe says: "We were kind of hinting to people the possibility of this neighborhood having some kind of continuity between what was and what's possible in the future." Although the buildings themselves are much larger than the shotgun houses preserved by Project Row Houses, their architecture echoes that of the historic housing and maintains its affordability. In a neighborhood where real estate went for "a dollar a square foot or less" in the early 1990s, by 2015, it was more than $60.00 per square foot. Lowe credits the city's over $30 million plan for the redevelopment of nearby Emancipation Park—Texas's first public park, founded by four freed slaves and at one time the only park in the city open to African Americans—with "sending signals out to developers around the city

looking to make a quick dollar."[82] The redevelopment is a major public investment in the neighborhood.

Lowe and his mostly faith-based partners in the newly established Emancipation Economic Development Council are now trying to figure out what to do next. In addition to a long-standing partnership with Rice University, Lowe also connected with the Department of Urban Studies and Planning at the Massachusetts Institute of Technology as well as with groups at nearby Texas Southern University and the University of Houston to help him figure out how to stifle gentrification in the neighborhood. One major effort has been strategic property acquisition. For example, one portion of the plan involves preserving commercial real estate on major intersections in a place without zoning and purchasing vacant lots in the middle of blocks so as to keep developers from being able to launch full-block development projects. Of all this protective purchasing, Lowe says: "Hopefully that's not pushing the market even higher."[83]

Lowe's work with Project Row Houses, along with the efforts of individuals and organizations such as Dennis Maher, Theaster Gates, and Big Car, highlights the potential of a form of arts-based placemaking that moves beyond recruiting and catering to the creative class. In very real ways, these actors have created spaces where new narratives can be produced, narratives that bring to the surface histories, identities, cultures, and ideas usually ignored by mainstream creative-placemaking endeavors. In such work, the arts are more than a catalyst for economic development or a tool for the recruitment of young professionals. Instead, they are a component of a broader, interconnected campaign to develop more equitable approaches to employment (for artists and nonartists alike), housing, and even urban planning in the early twenty-first-century city, all of which speaks to the culture, histories, and needs of established residents.

Yet Lowe's apprehension signals some of the other essential questions that creative placemaking raises but that are not often investigated, and his concerns illustrate the political and developmental contexts, complications, and tensions inherent in all placemaking efforts, particularly when it comes to matters related to race and class. For example, how effective can small-scale, arts-based programs be at addressing structural problems like gentrification or racism, even with the best of programs and intentions? Do placemakers who eschew or cannot gain government engagement thereby reinforce the neoliberal turn to using privately funded art to attract more and more private investment? And what happens when government officials begin to pay attention once such projects enjoy a modicum of success?

It is easy to ask such questions of a project, like Project Row Houses, that has been in existence for over twenty-five years and garnered attention from such high-profile funders as the MacArthur Foundation. In this book, we have suggested that the dominant conception of cre-

ative placemaking has often served to reinforce existing inequalities, and we have argued that there are ways to approach this practice that can mitigate such damage. This does not, however, suggest that such alternative approaches are not without similar pitfalls, particularly as they succeed and garner more attention (and subsequently more funding). At the same time, many of the projects outlined in this book had—unlike Project Row Houses—a limited life span, whether by design or owing to other factors. What is the impact of such ephemerality on the communities in which these projects were anchored? Does an expanded understanding of creative placemaking help us make sense of these issues or serve only to muddy the waters further? How, in other words, should we ultimately judge the success and failure of such projects? We turn to these questions in the conclusion.

Placemaking Is for People

(*Previous spread*) Boys bicycle in Braddock, Pennsylvania, with the Edgar Thomson Steel Works in the background.

This book has sought to present two things: a history of the rise of contemporary creative placemaking and case studies of projects offering a counterapproach. As such, the goal has not been to provide a definitive, unified, or comprehensive how-to guide to creative placemaking. Instead, it has been to decenter the discourse and demonstrate how the overlooked actors, ideas, and histories that inform such alternative approaches produce both tangible and intangible outcomes that can serve as a foundation for a fuller understanding of placemaking and its potential. Yet there may be no such thing as an ideal example of creative placemaking, and it is in that ambiguity that one can see both the promise and the peril of placemaking as an urban redevelopment tool. In other words, no single approach to placemaking is guaranteed to succeed, and no approach is destined always to fail. As we saw in Detroit and elsewhere, ArtPlace America, perhaps the standard-bearer of mainstream creative placemaking, has funded a number of projects that have brought real services and resources to communities in need of them. At the same time, projects that pushed (and push) placemaking's boundaries by incorporating racial and class understanding into their projects, like the Houston-based Project Row Houses, run the risk of becoming victims of their own success. As more homes are rehabilitated and constructed in the city's Third Ward, property values will inevitably continue rising, displacing the original residents.

New development in Project Row Houses' Third Ward.

Using two brief case studies, we attempt here to address such important matters. What these examples suggest is that both these approaches may be producing underappreciated and, in some cases, unintentional outcomes, and an understanding of such outcomes may allow us to suggest that it is time to stretch the definition of *placemaking*—if we even want to continue using such a word. For we may be doing more than asking the wrong questions to assess the outcomes of such work. We may even be using the wrong vocabulary. After all, emphasizing the parameters of creative placemaking misses the point about what needs to be done: constructing a process that produces the outcomes communities desire.

The Rise of "the Mayor of Rust": Creative Placemaking in Braddock, Pennsylvania

If one is looking for one city where all these narratives come together, where a grassroots approach to placemaking has gone mainstream, one could do worse than focusing on Braddock, Pennsylvania. Even though it is small—fewer than three thousand residents call the city home—it has become a national model of how placemaking can work. At the center of this story is Mayor John Fetterman, a six-foot-eight tattooed Harvard graduate. With his shaved head and goatee, Fetterman has embraced placemaking as a way to bring a Rust Belt city back to life.

In this city hit hard by capital mobility and deindustrialization, Fetterman is clear on how he sees placemaking as working: "We use art to combat the dark side of capitalism." More specifically, he notes: "We created the first art gallery in the four-town region, with artists' studios. We did public art installations. And, I don't know if you consider it arts, exactly, but I consider growing organic vegetables in the shadow of a steel mill an art, and that has attracted homesteading."[1] Others have taken note of the man the *New York Times* calls "the Mayor of Rust" and his "do-it-yourself aesthetic." He has spoken at such high-profile events as the Aspen Ideas Festival,

where he was introduced by former chairman of the National Endowment for the Arts Dana Gioia, appeared on such television programs as *The Colbert Report*, and been featured in such publications as *The Atlantic* and *Rolling Stone*. After losing a race for a US Senate seat in 2016, Fetterman made the decision to vie for the Pennsylvania lieutenant governor's office in 2018, successfully running on what he had done for Braddock. And his work in the troubled city has garnered him some big-name endorsements. Former governor Ed Rendell proclaimed, for example: "[Fetterman] revitalized the downtown, reduced the crime record.... Most of all, he brought hope back to a fine heritage Pennsylvania city in Braddock."[2]

Fetterman has embraced the arts in his efforts to revive Braddock. In 2009, he used family money to found a nonprofit organization, Braddock Redux, as a means of purchasing property and implement programming without getting caught in the red tape that often goes hand in hand with urban redevelopment projects. One Braddock Redux project is UnSmoke Systems Artspace, housed in a former Catholic school building that has been repurposed to feature a gallery, an events venue, and artist studio spaces. According to the space's website: "The project seeks to generate positive and intrepid ideas about the reuse of urban space. In a town where dilapidation and neglect have scarred the landscape, UnSmoke Systems contends that Braddock is fertile ground for creativity." UnSmoke also provides space to Braddock Avenue Books, an independent publisher, and Bibliopolis, a Berkeley-based company that offers e-commerce website development, technical support, and database services for antiquarian booksellers.[3]

Across the city, other entities are embracing similar strategies. The Braddock Carnegie Library—the first Carnegie library in the nation—runs a screen-printing program as well as a ceramics program. It is also home to Transformazium, an artist collaborative founded by the New York City transplants Ruthie Stringer, Dana Bishop-Root, and Leslie Stem. Transformazium works to provide

The Carnegie Free Library of Braddock.

Braddock Farms and the United
States Steel Edgar Thomson Plant.

opportunities and resources for artists who live in Braddock. In 2016, for example, it secured $35,000 through the Pittsburgh Foundation to support a residency for the multimedia artist and Braddock resident Tyrone Brown. Brown, who had been in and out of jail over the previous several years before finding stability through another program at the library, drew from his experiences to produce an audiobook of Nietzsche's *Ecco Homo* and a print book of writing and photographs. Here, critical attention is brought to an often-overlooked problem—what to do with individuals after they are released from detention facilities—but it is rebranded as art, not a state-sanctioned (and -funded) social service. Other Braddock-based residency programs include Into the Furnace, a writer-in-residence program that offers recipients up to nine months in the city to live and work.[4]

Like other cities wrestling with a glut of vacant spaces, Braddock has also embraced urban agriculture. The most prominent such project is Braddock Farms, which sits on an acre of land in the center of the city's business corridor. The farm was the result of a 2007 conversation between Mayor Fetterman and Grow Pittsburgh, a group that advocates for garden and food access. The farm was founded in the summer of that year and expanded in 2010, coming to include two hoop houses, a greenhouse, and a shipping container/office storage space.[5]

Young people come to work at the farm through an apprenticeship program and through the Braddock Youth Project, a youth employment program begun by Mayor Fetterman in 2006. The farm sells its produce through a farm stand in Braddock and the Penn's Corner Farmer Alliance, a farmer-owned cooperative operating in southwestern Pennsylvania that runs a community-supported agriculture program and supplies fresh produce to such restaurants as Superior Motors, a *Food and Wine* magazine "Restaurant of the Year" for 2018. Superior Motors takes it a step further and offers its own job-training program, one that is free of change and that, as the restaurant proclaims on its website, seeks to "introduce partic-

ipants to the urban agriculture and world class culinary/restaurant experience and skills necessary to excel in these fields."[6]

Other vacant lots have been repurposed in similarly innovative ways. In 2012, Gisele Fetterman—a Brazilian immigrant and the mayor's wife—opened Free Store 15104. Made up of three shipping containers on a previously vacant lot, the volunteer-staffed store collects goods from retailers and other donors and gives them away to families in need. It provides close to sixteen hundred families a month such things as clothing, furniture, household items, and toys.[7]

Numerous extant structures have also been repurposed. In May 2015, for example, a ribbon cutting was held for the Free Press Building. This building, which had once been home to the *Braddock Free Press* newspaper, Guentert's Bakery, and the Braddock Moose Lodge, had been vacant for close to a decade. New tenants included Studebaker Metals (a traditional metalsmithing workshop), a variety store, a pizza parlor, and a coffee shop, along with seven apartments on the upper floors.[8]

Such redevelopment seemed to culminate in August 2018 with the opening of the Braddock Civic Plaza. The plaza was built on the site of the former University of Pittsburgh Medical Center's Braddock facility, which closed its doors in 2010 and whose 2011 demolition, according to Mayor Fetterman, "left a giant gaping hole in the ground." The football field–sized plaza, located in the center of the city, has space for community gatherings and is lined with trees and flowers. Tables allow space for guests to eat food purchased from nearby food trucks. A circular event space can be found at the plaza's center. At the plaza's opening, Mayor Fetterman noted: "This community plaza represents the full, full circle in my mind. The entire site has not only been repurposed, but repurposed in the highest, best use possible."[9]

For Fetterman, this work was integral to restoring life in Braddock. In such cities, gentrification simply was not an issue the way it was in places like New York and San

Francisco. "Whether it's Detroit, or whether it's Buffalo, or whether it's Akron or Dayton, Ohio," Fetterman explained in an address given at the 2013 Creative Time Summit (a self-described "annual convening for thinkers, dreamers, and doers working at the intersection of art and politics"), "there's a lot more of us than there are of the beautiful cities." While cities adept at attracting the creative class may have to worry about becoming too popular and thereby pricing certain people out, Braddock actually suffered from the opposite problem. There, according to Fetterman, the main issue was not gentrification. It was "abandonment."[10]

In real ways, the logic proved compelling, and it seemed to make Braddock a safer, more livable city. In May 2013, for example, the city celebrated five years without a murder. Fetterman noted the importance of the Braddock Youth Project to such a development. Soon, migrants did start coming to Braddock, as did newspapers, magazines, and corporations. The city appeared to be going through a renaissance, and placemaking seemed to be at the heart of this rebirth.[11]

But those who had lived in Braddock for decades understood that this was not the first time the city claimed to be reborn. According to the local filmmaker Tony Buba, a critic of Mayor Fetterman's: "In the late '70s, a group of activists got together to reopen the abandoned Carnegie Library, and they had a lot of great programs out of there too. It was all part of a time here, when a lot of the socialists from the '60s ... were getting jobs in the steel mills, getting involved in the union." Such efforts, according to Buba, sought to engage with those already in Braddock and, more specifically, the region's African American community. To Buba, the new residents drawn to the cult of personality associated with Fetterman seemed to see Braddock as a blank slate. "Honestly," he concludes, "if you compare what these kids are doing to that last wave of activism ... [t]he group that did the library had a lot more buy-in from the black community in Braddock. . . . There was an energy from the civil rights movement and

everything else. I don't know that I see that these days."[12]

Instead, Fetterman and other newcomers have proved adroit at capturing the attention of those from outside the city. The most dramatic example of this occurred in 2010 when Levi's awarded Braddock $1 million to film a series of commercials in the city using Braddock residents as models. Yet the money from the jeans company did not go into city accounts; instead, it went to support Fetterman's own nonprofit organization, Braddock Redux. By 2011, Braddock Redux owned a former convent, a school, and a church, the latter being the home of the Braddock Youth Project. Such a strategy has allowed Fetterman to circumvent the standard democratic channels through which such urban redevelopment takes place. According to Braddock resident Pat Morgan, "Fetterman doesn't play well with others. He decides we need a youth center, and he's going to put it here, but I never hear him come to one of these council meetings and say, 'You know what, I've got some money, what should we do with it' If you're casting yourself as the mayor who speaks for Braddock, we want him to at least pretend that we have a say in any of this."[13]

Other critics of the Levi's campaign, which had been created by the firm Wieden + Kennedy Portland, called out the content of the television and print ads themselves. Working under the broader theme "Go Forth to Work," one television spot first focused on Braddock residents working to fix up dilapidated buildings, then continued with a voiceover: "A long time ago things got broken here. People got sad and left. Maybe the world breaks on purpose so we can have work to do."[14]

Such a conclusion from a multinational clothing manufacturer obviously overlooks the long and troubled history of deindustrialization in cities such as Braddock. Global forces within the steel industry led to the decline of Braddock, not the need for hipsters to have something to do. Yet for the photographer Latoya Ruby Frazier—whose family goes back four generations in Braddock—the Levi's campaign also highlighted how the popular narrative erased the practical realities and the structural forces at play in Braddock. Her voice was one of a chorus critiquing the advertisements along with similar campaigns elsewhere, like Palladium Boots' Johnny Knoxville–led "Detroit Lives" or the later Chrysler "Imported from Detroit" spots.

Now a professor of photography at the Art Institute of Chicago, Frazier made her name through "Campaign for Braddock Hospital (Save Our Community Hospital)." The series comments on the borough's current condition by intertwining photographs of derelict Braddock buildings with criticism of the Levi's advertising campaign and the aforementioned University of Pittsburgh Medical Center decision to shutter the Braddock General Hospital. Frazier critiques both institutions through her manipulation of ads and captioning of photographs, asking: "Do we really need $50 jeans or $250 trucker jackets when we don't have medical care?" This challenge also spills over to Levi's beneficiaries, including local arts spaces and nearby towns, simultaneously attacking capital's appropriation of grassroots creative production and the geographic winners and losers of deindustrialization. When the city's only hospital was shuttered, it was the latest example of the decades-long abandonment of the borough by those outside it, and the Levi's ads symbolized how Fetterman's promise for the future of Braddock could celebrate a future for the city wrapped up with the aesthetics of its history rather than addressing the persistence of race and class conflict that produced it. In real ways, the placemaking campaigns of the twenty-first century are taking advantage of these longer narratives of abandonment—often without realizing their complicated nature or even their existence. While Frazier and Fetterman both attacked the closure of the medical center, Frazier's series implicates Fetterman's strategy for growth as a new kind of extractive capitalism. For every program that addresses the consequences of neoliberalism and structural racism by providing real support for residents like Tyrone Brown, there is another that seems to do more for the companies or newcomers than for the families who have called Braddock home for decades.[15]

Such dual realities suggest that the Braddock situation is not as clear-cut as Fetterman's critics and supporters suggest. The work spearheaded by the former mayor has raised unprecedented awareness of and funding sorely needed by similarly troubled communities around the country. And much has been done in Braddock, from redeveloping the former hospital site to implementing programs that improve access to services and opportunities for those who need them the most. But the organizational framework used to guide the bulk of this funding, particularly as directed through Braddock Redux, was nowhere near the inclusive, democratically designed effort it could have been. One result of this shortcoming is that some activists—particularly those who feel left out of Fetterman's process—are mobilizing around key questions. For whom are such efforts being made? Who is empowered by these projects? And who is left out of the development process?

To the Braddock community activist Isaac Bunn, Fetterman's efforts have privileged white newcomers while ignoring the voices of residents of color, many of whom have for generations called the city home. This belief has led Bunn to refer to such an approach to placemaking as "placelessness-making." Discussing the economic-redevelopment strategy employed by Fetterman and his allies, Bunn notes that the mayor's nonprofit organization, Braddock Redux, often takes advantage of its "direct privileged hotline" to local foundations, bypassing the standard channels of democratic debate and decisionmaking. The result has been that the city's residents—and its African American population in particular—have been left out of the discussion and are therefore less likely to benefit from this brand of placemaking. Conversely, those who already possess financial and social capital are placed in positions from which they may disproportionately gain from such developments. "His [Mayor Fetterman's] activities betray the democratic ideal of having an equitable, inclusive, and just civil society," comments Bunn. "Is inclusive social imagination at work in creative-placemaking activities when enclaves of privilege are developed in which the benchmark of success is a 'high-end' restaurant [Superior Motors] placed in one of Pennsylvania's poorest communities?"[16]

In answering this query with a definitive "no," Bunn is ultimately rejecting the placemaking strategy of Fetterman. Even considering the much-needed tax revenues and job-training opportunities provided by the likes of Superior Motors, is the restaurant what would have emerged from a different development strategy? To Bunn's mind, such projects are illustrative of "the blind love of creative placemaking that is tied to the allure of outsider speculation culture and its economic thinking of 'build it and they will come.'" This approach "supports a politics of disbelonging employed to manufacture a privileged 'playground' place." A new approach to placemaking is required.[17]

Bunn sees the seeds of such an alternative—as well as one outcome of such critiques—in the political campaign of Summer Lee, who in May 2018—with the backing of the Democratic Socialists of America—won the Democratic primary for state representative in Allegheny County. On January 1, 2019, she became the first African American woman to represent southwestern Pennsylvania in the state legislature. Lee was inspired by Frazier's work against the Levi's ad campaign, which she has called "a Rust Belt whitewash," and she considered Frazier "a Black woman from Braddock speaking truth to power." Importantly, Lee also ran on a critique of what Mayor Fetterman had been doing in Braddock. She defeated the incumbent and centrist Paul Costa, who was endorsed by Fetterman and touted such things as Superior Motors as central to the revitalization of Braddock. Lee—unlike Fetterman—regarded such development as an early phase of gentrification, one that would she worried lead to another round of displacement of African Americans in the region.[18]

Bunn volunteered for the Lee campaign and has called on her to embrace an "aesthetic of 'belonging' as central to future placemaking [efforts]." In Lee, Bunn sees a local politician who will work on placemaking projects that call for "economic, social, cultural, and civic belonging—how to create it; how to understand and accommodate cultural difference in matters of civic participation; how to enhance the community's understanding of citizenship beyond the confines of leisure pursuits and consumption; how to help the citizens of a place achieve strength and prosperity through equity, economic opportunity, technology educational advancement, social structural change programming and civility."[19] Her platform seems to suggest that she is taking such advice to heart. She has argued for universal free prekindergarten, a moratorium on fracking (Fetterman has supported some fracking as part of a job-creation strategy), full state funding of lead-waterline replacement, a $15.00-per-hour minimum wage, a single-payer health care system, the elimination of cash bail, and an opposition to the reinstatement of mandatory minimum sentencing laws. As she said at a Pittsburgh City Council meeting about a hospital expansion: "'Revitalization' doesn't help people. . . . They need better wages." While this approach is undoubtedly place based, it is far from a strategy meant to excite the creative class, and it was not necessarily even perceived as an outcome of Braddock's creative-placemaking strategy. It may, however, show what could come to happen in cities across the United States as residents grapple with the positive and negative outcomes of creative-placemaking agendas.[20]

Growing People: The Evolution of Growing Power in Milwaukee, Wisconsin

In many ways, the example of Braddock and Fetterman shows the unintended consequences of a successful placemaking campaign. But what happens when such an endeavor seems to fail? In November 2017, Growing Power—the urban farm founded by Will Allen in Milwaukee—abruptly closed. That month, the organization's board of directors voted to dissolve, citing Growing Power's financial problems, including onetime operating deficits of more than $2 million and pending judgments for more than $500,000. It was a sad end to an organization on the cusp of its twenty-fifth year in operation.[21]

The organization's undoing may have been caused by its success following Allen's prestigious MacArthur award in 2008. What was already a quickly growing group began developing at an unsustainable rate. Among the numerous opportunities offered at the time were major expansions of the organization's operations, including through a W. K. Kellogg Foundation $5 million grant to develop community food centers across the United States. As Growing Power's projects swelled and employed more than 150 people, the board remained relatively hands-off, with little midlevel management or organizational structure to support Allen, who remained the center of the organization.[22]

Others came to question the politics of the organization. In 2011, Growing Power accepted a $1 million grant from the Walmart Foundation. Some critics saw this as an implicit endorsement of Walmart's overall practice of underpaying farmers and other food-chain workers. As Andy Fisher, the cofounder of the Community Food Security Coalition, noted: "I thought it was naïve and problematic that he was taking the money and giving them a pass on their payment practices."[23]

Yet, even with these complications, Ricardo Salvador, the director of the food and environment program at the Union of Concerned Scientists and a former Kellogg Foundation program officer, assessed: "The training, learning, and benefits of Growing Power will be felt for years to come." There is little doubt that he is correct. In March 2018, Allen's daughter, Erika, reorganized the Chicago branch of Growing Power into the Urban Growers Collective (UGC). UGC eschews high-production farming for an emphasis on empowering young people of color through educational programming, job-training apprenticeships, and leadership training. And, on an even larger level, Allen inspired thousands of individuals to start their own farms—or even just take control of their lives. As the online publication *Civil Eats* describes this phenomenon: "It's clear that Will Allen's legacy will live on in the many organizations that grew from his work. In addition to em-powering a generation of community leaders all across the country, who have gone on to radically transform their lives and neighborhoods, he also succeeded at teaching and protecting vulnerable Black children in an era when very few other entities were up to the task."[24]

On the macro-level, this has led to the creation of numerous new spaces and many new farms. For example, from 2009 through 2017, in locations in such states as New Jersey, Colorado, Massachusetts, Arkansas, Kentucky, Georgia, and Mississippi, Growing Power Regional Outreach Training Centers developed sites that improved access to both healthy food and economic opportunities for farmers and residents alike. Yet what is most remarkable is that others are building on such work, as is evident in what the sociologist Monica M. White describes as "the blossoming and expansion of the current African American urban agricultural movement" across the United States. Farmers such as the Philadelphia-based Chris Bolden-Newsome see such spaces as sites of reconciliation as well as production. "As Black farmers," he notes, "we need to be more than just growers of food. We have to be teachers of culture and healers of trauma. That is how we survive." In Atlanta, the Grow Where You Are Collective also uses agriculture as a way to address the overall health of the city's African American population: the farm regularly offers free meals, health-related workshops, and lessons in capoeira (an Afro-Brazilian martial art form). Such a site of activity, the founder, Eugene Cook, says, "[illustrates] what a sovereign community can feel and look like."[25]

Back in Milwaukee, the example of Dasia Harmon illustrates what such a community can produce. Harmon grew up on the North Side of Milwaukee, and, in 2001, when she was nine years-old, her mother heard about Growing Power and thought it would be "a cool place for [Dasia] to spend a summer." Harmon was not thrilled: "I remember vividly that I didn't want to be there." The sight of produce growing and the smells of composting

Growing Power's urban farm before it transitioned to Will's Roadside
Farm and Markets.

and farm animals were foreign to a city kid like her. Yet, despite the rooted nature of the farm experience, she recalls Growing Power instilling in her a sense of mobility. Even as a child, she was given the opportunity to travel around the metropolitan region representing the urban farm at a variety of farmers' markets. There, she saw just how influential Allen and Growing Power were. "I learned," she recalls, "how much of a difference it [Growing Power] made in the community." And, as she spent more time working with Allen, she came to find the smells and sights of the farm comforting and soothing. The farm, in other words, became a place where she felt safe and where she wanted to spend time.[26]

When Harmon was a teenager, a group of representatives from Fort Valley State University—a historically Black land-grant university in Georgia—came to a workshop at Growing Power. They were so impressed with her they offered her a scholarship on the spot; she went on to earn a bachelor of science in agriculture from the school. The city kid who once found the sights and smells of a farm discomforting now wanted to be academically immersed in such things. Wanting to learn more about agriculture, Harmon next earned a master's degree in entomology from Florida Agricultural and Mechanical University. Fresh out of graduate school, she moved to the Greater Atlanta area and began working with Food-Corps, an organization started in 2010 with the intent of connecting children to healthy food in school. As it turns out, Will Allen was instrumental in the founding of FoodCorps; the first cohort of fifty corps members trained with Allen at Growing Power before fanning out to ten states across the country. For Harmon, however, it was at FoodCorps that she discovered her "passion for teaching."[27]

In October 2017, Harmon took a position as a health educator with HealthMPowers, Inc., an Atlanta-based nonprofit organization that works to promote healthy eating and physical activity in schools. It is now her job to develop curricula centered on cooking and gardening for twenty-four Boys and Girls Clubs housed in public schools in the metropolitan Atlanta region. Commenting on this position, she notes: "I really feel like I'm making a difference." And it all began at Growing Power. "Will Allen and Growing Power," Harmon concludes, "played a very significant role in my life."[28]

By the summer of 2018, Allen had launched a new, smaller venture, Will's Roadside Farm and Markets, the name being a nod to his pre–Growing Power days when he sold his produce to anyone who would buy it. As he did in the past, he continued to offer market baskets of assorted produce, organic eggs, and other traditional farm products to customers. But he has also used this moment of realignment to experiment with hemp cultivation and cannabidiol (CBD) oil production. Such products can be used to alleviate both physical and mental pain, but they also provide a revenue stream. This new endeavor is organized as a for-profit company as Allen now sees the need to move away from the nonprofit model for such work. "Nonprofits," he notes, "have to get soft money to operate and you have to have cash flow for a farm to survive. It cannot be done by a nonprofit." Yet, despite the emphasis on the project's for-profit status, there is still something place based in Allen's desire to continue farming. "I never did this for the money," Allen noted in June 2018. "It is therapy for me to touch the soil every day. I tell that to people who are suffering. Go touch the soil." Despite the changes to his organizational model, he remains committed to farming at his Fifty-Fifth and Silver Spring site. This was his community, his home.[29]

Learning from Success and Failure

Despite the complications of Growing Power's and Braddock's experiences, creative placemaking remains standard practice for a wide variety of American cities; this will not change in the near future. The successes and/or failures of each case are indicative of those of the hundreds of ventures we visited and followed throughout this project. Many of these problems are familiar to not-for-profit workers and scholars, including overemphasizing and overrelying on individual leaders and having goals that are difficult to articulate or assess. We are already aware of some solutions to such problems, including building coalitions that distribute leadership and responsibility, alongside making space for more comprehensive, representative, and articulated visions for success. Established community groups and foundations are starting to bring this understanding into creative placemaking, but there is still a long way to go, particularly for organizations at either end of the spectrum: projects led by established organizations doing business as usual and ephemeral and underfunded projects disconnected from external funding and support networks. The second category of groups—those outside the mainstream—also includes those pushing creative placemaking toward new problems and new solutions. They—and we—ask, How can we best ensure that sociability is infused with productive potential? How can we ensure that those marginalized by racism, economic inequality, and other systemic powers receive the support needed to make sure their projects achieve their full potential, especially when those projects are not the kinds supported by typical funders? In many cases, highlighting the fact that such projects do not exist in a vacuum—indeed, that there is an informal network doing similar work across the country—is a valuable first step. Moreover, there is power in collective action here, in smaller actors banding together to strengthen their potential. One sees this in such endeavors as The Plant in Chicago and Ponyride in Detroit. We must also learn from those projects, such as Big Car in Indianapolis, that continually reexamine their role in changing the face of a neighborhood, for better and worse.

Finding answers to these questions becomes ever more urgent as the federal government continues to step further away from the funding process, as is evident in the sunsetting of the entire ArtPlace America program in 2020. This means that it will be incumbent on states and municipalities to shoulder the full load—and to do more with less. On the one hand, such a moment of transition is undoubtedly unsettling. How will placemaking function without access to federal money? Yet it could also offer opportunities for experimentation, allowing the notion of placemaking to be taken in new and exciting directions. The cases of SOUP in Detroit and FEAST in New York City illustrate what alternative methods of raising much-need capital could look like.

This is one reason exploring fundamental changes—or alternatives—to creative placemaking is so important; they could serve as a sort of road map for such an evolution. At the heart of this effort should be a rethinking of the relationship between place and production. As the Braddock example shows, this often means art when it comes to placemaking. But, as examples like The Work Office in New York City, the programs of Project Row Houses in Houston, and even the design arts training provided by Assembly House 150 in Buffalo all suggest, using the arts to create new senses of place can go beyond installing murals and offering artist residencies (though there is nothing wrong with either of those things): they can provide such valuable things as social capital, housing, and job training and opportunities.

Yet, as Will Allen continues to show, this can also mean the production of actual things and the economic arrangements and opportunities that such items entail. This can be healthy food, as with myriad urban farms that now dot the landscape of cities like Milwaukee, where groups like Walnut Way have used urban agriculture as a spur to more place-based economic redevelopment. It can even be wine, as with Cleveland's Chateau Hough, and potato chips, as with Detroit's Detroit Friends. It can even be light manufacturing, as with Detroit's nonprofit Empowerment Plan hiring homeless women to manufacture its innovative hybrid coat/sleeping bag along with the line of clothing offered through its Maxwell Detroit retail company. The growth of the Empowerment Plan, along with Will Allen's continued evolution, perhaps suggests that we may need a better understanding of the role of for-profit endeavors in the placemaking discussion and of their relationship to nonprofit organizations. This is particularly the case as grant opportunities and government funding for such work continue to disappear.

Yet, as the cases of both Summer Lee and Dasia Harmon suggest, we may want to focus on other types of production. How can place produce new political and professional opportunities, particularly for those who, like African American women, are often left out of mainstream placemaking endeavors? For example, Braddock's realities were a major reason Lee was compelled to use the political party system to address issues that some have seen creative placemaking as addressing. Moreover, the experience of Growing Power led Harmon to a successful career teaching public school students in Atlanta. Neither of these outcomes would have happened without a deep engagement with the community. Yet the type of work both women are doing involves growing the very institutions that the neoliberal city has weakened and that placemaking efforts of all sorts often overlook: political parties and public education systems. Here are hints of a strategy that can address the shortcomings that placemaking is perceived as speaking to through more democratic and inclusive means. Indeed, mainstream placemaking is often posited as a counter to formal politics and education systems, which are both seen as dysfunctional, if not outright broken. A move to strengthen such institutions—and to make them more responsive to a broader variety of urban residents—will allow for both a more robust public sphere and healthier urban communities in ways that mainstream placemaking simply cannot deliver.

With both insider and outsider creative-placemaking efforts, the placemaking process often starts with the simple question, How could we improve this place? Commonly, that question is asked about a derelict lot or an underused building, in which case almost anything seems like an improvement. After all, who could argue with cleaning up the trash and putting *something* there? But the simplicity of the question hides how complicated it is to define *improvement* and *improvement for whom*. Did Fetterman and placemaking in Braddock succeed? Go ask Summer Lee. Did Will Allen and Growing Power fail? Go ask Dasia Harmon. As did the activists who fought for the La Casita library in Chicago, Lee and Harmon began their work by engaging with efforts regarding the physical redevelopment of urban spaces. Yet, over time, both saw how such efforts could lead to the creation of more intangible assets, whether they be in the arena of political networking (Lee) or curricula development (Harmon). And there is the goal that such intangibles will serve as tools to produce new, more equitable spaces and cities. But, for both, as for the Pilsen activists, there is the hope that the development of such intangible assets will have a direct impact on the built environment of the cities they call home. State funding for lead-water-line replacement means revitalized housing in Braddock, while the adoption of Harmon's curricula in Atlanta means more gardens for the city's school-age children.

We call for an approach to placemaking that recognizes these relationships and complexities. At present, creative placemaking is all too often about someone's idea of the social aspect of a place, but, by overemphasizing sociability, we underemphasize the other aspects of being human. In other words, what may be required is a strategy of direct political engagement coupled with a more robust production-oriented placemaking, both of which are fueled by the need to build new relationships, recognize voices, and act. Such a strategy can lead to increased social capital, but in a way that benefits already-established residents. It is not a strategy meant to entice the creative class.

Determining which actions or outcomes are appropriate requires understanding the collective and personal histories that inform such developments. These histories, usually not a part of the contemporary-placemaking discussion, matter. They are the histories of individuals and movements often left out of the placemaking discourse. What we see here is that they can be powerful when drawn to the surface. In other words, we have to produce new understandings of the histories that inform placemaking. These histories must be attuned to the realities of race and class (and so much more) in ways mainstream placemaking has not been.

Placemaking is, indeed, about transforming place, but it is also about people, and it should be about all of us.

Acknowledgments

We would like to acknowledge and thank all the organizations, groups, and individuals that took the time to speak and work with us while we collaborated on this project over the last ten years. We truly appreciate your generosity.

We would also like to thank the institutions and organizations that have supported this project, including the Graham Foundation, the Milwaukee School of Engineering (Eric Baumgarter, vice president of academics, and Alicia Domack, Humanities, Social Science, and Humanities department chair, Bridgette Binzcak, DeAnna Leitzke, and Kelly Ottman), the Grohmann Museum (especially James Kieselburg, Ann Rice, and Russell Piant), St. Olaf College, and, at the University of Chicago Press, Tim Mennel, Doug Mitchell, Susannah Engstrom, Tyler McGaughey, Michael Koplow, and Joseph Brown.

David

I deeply appreciate the inspiration and support of my colleagues at the University of Chicago and St. Olaf College; the community of photographers, filmmakers, and other artists with whom I have collaborated; and my friends and family. Thank you.

Michael

I would like to thank Neil Harris, George Chauncey, Kathleen Conzen, and Adam Green, who taught me what it means to be a curious, conscientious scholar. I am also thankful for Lizabeth Cohen, Jamin Rowan, Joseph Heathcott, Andrew Karhl, Nathan Connolly, Amanda Seligman, Nicolas Lampert, Paul Kjelland, Raoul Deal, Molly Hudgens, David Spatz, Brian Goldstein, Jennifer Hock, Robert Smith, Virginia Small, LaDale Winling, Meredith TenHoor, Susan Sloan, Joan Ockman, Dave Boucher, Arijit Sen, and Samuel Zipp. Each of you, whether you know it or not, has compelled me to think about urbanism in different ways. And a special shout-out to Michael Stamm and Mike Czaplicki: thank you for over twenty years of friendship and support.

Outside the academy, I would like to thank Matt Mannherz, Nate Flanigan, Rich Leiter, and Dave Moore. Traveling the country in a van with the four of you gave me the opportunity to fall in love with countless American cities. I also thank my parents, Armand and Ellen Carriere, for so many years of support. Finally, I dedicate this book to my immediate family: Aidan, Liam, Esther, and Shelly. I look forward to exploring more cities with the four of you. I love you with all my heart.

Notes

Introduction

1. Richard Feloni, "Billionaire Dan Gilbert Has Already Bet $5.6 Billion on Detroit's Future, but Money Can't Solve His Biggest Challenge," *Business Insider*, August 18, 2018, https://www.businessinsider.com/quicken-loans-dan-gilbert-detroit-2018-8; Kirk Pinho, "Gilbert Real Estate Portfolio at 61 Percent by 2022 . . . Perhaps," *Crain's Detroit Business*, December 21, 2017, https://www.crainsdetroit.com/article/20171221/blog016/648251/gilbert-real-estate-portfolio-growth-envisioned-at-61-percent-by.
2. "Our Mission," n.d., Project for Public Spaces, https://www.pps.org/about; "What Is Placemaking?," n.d., Project for Public Spaces, https://www.pps.org/article/what-is-placemaking; David Muller, "4 Keys to Dan Gilbert and Planners' Vision for Successful 'Placemaking' in Detroit," *Detroit Business News*, April 11, 2013, https://www.mlive.com/business/detroit/index.ssf/2013/04/4_keys_to_dan_gilbert_and_plan.html.
3. "Opportunity Detroit: A Placemaking Vision for Downtown Detroit," Spring–Summer 2013, 5, 8, http://opportunitydetroit.com/wp-content/themes/Opportunity_Detroit/assets/PlacemakingBook-PDFSm.pdf.
4. Kirk Pinho, "Dan Gilbert Rolls Out a Vision for Detroit That Includes Districts, Connections, New Retail—and Papa Joe's," *Crain's Detroit Business*, March 28, 2013, https://www.crainsdetroit.com/article/20130328/NEWS/130329856/dan-gilbert-rolls-out-a-vision-for-detroit-that-includes-districts.
5. William H. Whyte, *The Social Life of Small Urban Spaces* (New York: Project for Public Spaces, 1980), quoted in "Opportunity Detroit," 10.
6. Richard Florida, *The Rise of the Creative Class . . . and How It's Transforming Work, Leisure, Community, and Everyday Life* (New York: Basic, 2002); Richard Florida, "How Detroit Is Rising," May 15, 2012, CityLab, https://www.citylab.com/life/2012/05/how-detroit-rising/1997.
7. Jane Jacobs, *The Death and Life of Great American Cities* (New York: Vintage, 1961; rev. ed., New York: Random House, 1993); John Galla-gher, "Urban Author Jane Jacobs' Influence Inspiring Detroit Projects," *Detroit Free Press*, December 3, 2016, https://www.freep.com/story/money/business/columnists/2016/12/03/jacobs-detroit-cities-planning-cox/94549440; "Hateraide: Why Downtown's Revitalization Requires Critics as Well as Boosters," *Deadline Detroit* (blog), January 9, 2014, http://www.deadlinedetroit.com/articles/7865/hateraide_why_downtown_s_revitalization_requires_critics_as_well_as_boosters; "Story," n.d., City Modern, https://www.citymoderndetroit.com/story.html (link inactive).
8. Dan Gilbert, "Detroit 2.0: It's Real and It's Happening Now," *Choose Thinking* (blog), July 17, 2011, https://choosethinking.com/2011/07/detroit-2-0-its-real-and-its-happening-now (link inactive).
9. See, e.g., "Garage Cultural: Funding Received," 2017, ArtPlace America, https://www.artplaceamerica.org/funded-projects/garage-cultural; and "Our Town: Grantees: Detroit Economic Growth Association," 2013, National Endowment for the Arts, https://www.arts.gov/national/our-town/grantee/2013/detroit-economic-growth-association. Garage Cultural, a Southwest Detroit group focusing on arts and cultural programming that received $250,000, is representative of such strategies.
10. Ann Markusen and Anne Gadwa, "Creative Placemaking" (Washington, DC: National Endowment for the Arts, 2010), 3, 10.
11. Susan Silberberg, "Places in the Making: How Placemaking Builds Places and Communities" (Cambridge, MA: Department of Urban Studies and Planning, Massachusetts Institute of Technology, 2013), 3, https://dusp.mit.edu/sites/dusp.mit.edu/files/attachments/project/mit-dusp-places-in-the-making.pdf; Jenna Moran, Jason Schupbach, Courtney Spearman, and Jennifer Reut, "Beyond the Building: Performing Arts and Transforming Place" (Washington, DC: National Endowment for the Arts, 2014), 6, https://www.arts.gov/sites/default/files/beyond-the-building-performing-arts-transforming-place.pdf.
12. Silberberg, "Places in the Making," 2, 6, 7.
13. At the city level, see, e.g., "The City of Atlanta Placemaking Program," n.d., City of Atlanta, GA, https://www.atlantaga.gov/government

/departments/city-planning/placemaking. At the state level, see, e.g., "Placemaking Indiana," n.d., Indiana Development and Community Housing Authority, https://www.in.gov/myihcda/placemakingindiana .htm. For foundations and placemaking, see, e.g., "Journey to Creative Placemaking: Lessons and Insights," n.d., Kresge Foundation, https:// kresge.org/content/journey-creative-placemaking-lessons-and -insights. For the private sector, see, e.g., "Southwest Airlines' Heart of the Community," n.d., Project for Public Spaces, https://www.pps.org /heart-of-the-community.

14. David Harvey, *The Enigma of Capital and the Crises of Capitalism* (New York: Oxford University Press, 2010); Mark Blyth, *Austerity: The History of a Dangerous Idea* (New York: Oxford University Press, 2013).

15. As Sassen writes: "And yet there are sites where it all comes together, where power becomes concrete and can be engaged, and where the oppressed are part of the social infrastructure for power. Global cities are one such site." Saskia Sassen, *Expulsions: Brutality and Complexity in the Global Economy* (Cambridge, MA: Harvard University Press, 2014), 11.

16. Markusen and Gadwa, "Creative Placemaking," 15; Ann Markusen, "How Cities Can Nurture Cultural Entrepreneurs," Ewing Marion Kauffman Foundation Policy Brief, November 2013, 2, https://www .kauffman.org/-/media/kauffman_org/research-reports-and-covers /2013/11/cultural-entrepreneurs-report.pdf.

17. Roberto Bedoya, "Creative Placemaking and the Politics of Belonging and Dis-Belonging," *GIA Reader* 24, no. 1 (Winter 2013), https:// www.giarts.org/article/placemaking-and-politics-belonging-and-dis -belonging.

18. For the impact of creative placemaking on urban economic redevelopment, see James A. Anderson, "How the Creative Placemaking Tide Lifts All Community Boats," June 10, 2019, Next City, https://next city.org/features/view/how-the-creative-placemaking-tide-lifts-all -community-boats.

19. Sharon Zukin, *The Cultures of Cities* (Cambridge, MA: Blackwell, 1995), 2.

20. Bedrock, "We screwed up badly the graphic package that was partially installed on the retail windows of the first floor of the Vinton Building …," Facebook, July 23, 2017, https://www.facebook.com /BedrockDetroit/posts/we-screwed-up-badly-the-graphic-package -that-was-partially-installed-on-the-reta/1528777023855661; Spivey quoted in Robert Allen, "Gilbert on Controversial Downtown Sign: 'We Screwed Up,'" *Detroit Free Press*, July 24, 2017, https://www .freep.com/story/news/local/michigan/detroit/2017/07/24/gilbert -downtown-detroit-sign/503929001.

21. John Gallagher, "Gilbert Dismisses Criticism of 500-Camera Security System," *Detroit Free Press*, August 16, 2015, https://www.freep. com/story/money/business/michigan/2015/08/15/gilbert-security- quicken-detroit-downtown-privacy/31635245; Jane Jacobs, *The Death and Life of Great American Cities* (New York: Vintage Books, 1961), 35.

22. "Quick Facts: Detroit City, Michigan," updated July 1, 2018, US Census Bureau, https://www.census.gov/quickfacts/fact/table/detroit citymichigan/PST045217. For representative job postings for such work, see "Securitas Jobs," n.d., https://www.careerbuilder.com/jobs -securitas.

23. Juell Stewart, "Chicago's Fight Over 'La Casita' Reveals Rifts in School Reform," *Colorlines*, September 24, 2010, https://www.color lines.com/articles/chicagos-fight-over-la-casita-reveals-rifts-school -reform.

24. Jiggetts, "Parents Build Library in Building Slated for Demo."

25. Joe Barrett, "Parents Enroll in Protest: Sit-In Stymies Chicago School's Plan to Demolish a Run-Down Community Hub," *Wall Street Journal*, October 1, 2010, https://www.wsj.com/articles/SB1000142 405274870478940457552418352819788.

26. Bob Roberts, "Makeshift Library Opens as Sit-In Continues," October 1, 2010, CBS 2 Chicago, https://chicago.cbslocal.com/2010 /10/01/makeshift-library-opens-as-school-parents-continue-sit-in.

27. Cinnamon Cooper, "The Kids Who Fight for a Library," September 28, 2010, *Gapers Block*, http://gapersblock.com/mechanics/2010 /09/28/the-kids-who-fight-for-a-library.

28. Ana Beatriz Cholo, "Little Village Getting School It Hungered For," *Chicago Tribune*, February 27, 2005, https://www.chicagotribune .com/news/ct-xpm-2005-02-27-0502270311-story.html.

29. Erik Gellman, "Pilsen," 2005, *Encyclopedia of Chicago*, http://www .encyclopedia.chicagohistory.org/pages/2477.html; Jessica Pupovac, "History of Pilsen," n.d., My Neighborhood Pilsen, WTTW Chicago, https://interactive.wttw.com/my-neighborhood/pilsen/history.

30. Jackie Kostek, "Whittier Field House Demolition 'Had to Happen,' Alderman Says," DNAinfo, August 20, 2013, https://www.dnainfo.com /chicago/20130820/pilsen/ald-solis-on-whittier-fieldhouse-this-had -happen.

31. David Schalliol, "Whittier Field House Library Ribbon Cutting Ceremony," October 1, 2010, *Gapers Block*, http://gapersblock.com /mechanics/2010/10/01/whittier-field-house-library-ribbon-cutting -ceremony.

32. Author interview with Lisa Angonese, July 9, 2018. For more on the history and efforts of PERRO, see https://pilsenperro.org.

33. Author interview with Lisa Angonese, July 9, 2018; Noreen S. Ahmed-Ullah, "Battle of All Mothers at Whittier Field House Draws

Close to an End," *Chicago Tribune*, October 20, 2010, https://www
.chicagotribune.com/news/ct-xpm-2010-10-20-ct-met-whittier
-cps-meeting-20101020-story.html; Schalliol, "Whittier Field House
Library Ribbon Cutting Ceremony."

34. Author interview with Nell Taylor, July 13, 2011; author interview
with Lisa Angonese, July 9, 2018.

35. Will Guzzardi, "Whittier Library: Parents, Police in Standoff Outside
Elementary School Fieldhouse," *Huffington Post*, June 23, 2011, https://
www.huffingtonpost.com/2011/06/23/whittier-library-parents
-_n_883540.html; author interview with Katherine Darnstadt, August
2, 2018.

36. Kati Gibson, "'They tore it down before our eyes!' CPS Contractor
Begins to Destroy La Casita Despite Library Treasures and Supposed
'Asbestos Danger,'" *Substance News*, August 17, 2013, http://www
.substancenews.net/articles.php?page=4449.

37. Author interview with Lisa Angonese, July 9, 2018.

38. Author interview with Lisa Angonese, July 9, 2018.

39. Author interview with Evelyn Santos, July 13, 2018.

40. Author interview with Nell Taylor, August 9, 2018.

41. Author interview with Katherine Darnstadt, August 2, 2018. For
more on Darnstadt's work, see Gilad Meron and Mia Scharphie, "How
to Become a Social Impact Designer without Going (Permanently)
Broke," January 16, 2017, Next City, https://nextcity.org/features/view
/latent-design-katherine-darnstadt-social-impact-architecture.

Chapter 1

1. Kenneth T. Jackson, *Crabgrass Frontier: The Suburbanization of the
United States* (New York: Oxford University Press, 1985), 238, 4.

2. Elaine Tyler May, *Homeward Bound: American Families in the Cold
War Era* (New York: Basic, 1988); David Riesman in collaboration with
Reuel Denney and Nathan Glazer, *The Lonely Crowd: A Study of the
Changing American Character* (New Haven, CT: Yale University Press,
1950); Robert Nisbet, *The Quest for Community: A Study in the Ethics
of Order and Freedom* (New York: Oxford University Press, 1953); Max
Lerner, *America as a Civilization* (New York: Simon & Schuster, 1957).

3. For a history of such ideas, see Steven Conn, *Americans against
the City: Anti-Urbanism in the Twentieth Century* (New York: Oxford
University Press, 2016).

4. Samuel Zipp, "The Roots and Routes of Urban Renewal," *Journal of
Urban History* 39, no. 3 (May 2013): 366–91.

5. William H. Whyte, *The Organization Man* (1956), rev. ed. (Philadelphia: University of Pennsylvania Press, 2002), 63, 68–69.

6. Whyte, *The Organization Man*, 289, 268.

7. Whyte, *The Organization Man*, 287, 289, 285, 300.

8. Whyte, *The Organization Man*, 9, 364.

9. Whyte, *The Organization Man*, 397.

10. Whyte, *The Organization Man*, 276.

11. Whyte, *The Organization Man*, 400, 403, 404.

12. Jacobs, *The Death and Life of Great American Cities*, 43, 506, 507
(emphasis in original).

13. Jacobs, *The Death and Life of Great American Cities*, 155–56.

14. Jacobs, *The Death and Life of Great American Cities*, 188, 516.

15. Jacobs, *The Death and Life of Great American Cities*, 188, 196, 197.

16. Jacobs, *The Death and Life of Great American Cities*, 511, 244.

17. Jacobs, *The Death and Life of Great American Cities*, 189, 191.

18. Jacobs, *The Death and Life of Great American Cities*, 117.

19. Jacobs, *The Death and Life of Great American Cities*, 112.

20. Jacobs, *The Death and Life of Great American Cities*, 153, 155, 159.

21. Jane Jacobs, *The Economy of Cities* (New York: Vintage, 1969), 6.

22. Jacobs, *The Economy of Cities*, 74, 75, 77.

23. Jacobs, *The Economy of Cities*, 245.

24. Jacobs, *The Economy of Cities*, 248.

25. Jacobs, *The Economy of Cities*, 95, 209, 210.

26. Jacobs, *The Economy of Cities*, 163.

27. Jacobs, *The Economy of Cities*, 249.

28. Jane Jacobs, *Cities and the Wealth of Nations: Principles of Economic Life* (New York: Random House, 1984), 5–6.

29. Jacobs, *Cities and the Wealth of Nations*, 31–32.

30. Jacobs, *Cities and the Wealth of Nations*, 32.

31. Jacobs, *Cities and the Wealth of Nations*, 221, 222.

32. Jacobs, *Cities and the Wealth of Nations*, 224, 222.

33. Jacobs, *Cities and the Wealth of Nations*, 230.

34. Kevin Lynch, *The Image of the City* (Cambridge, MA: MIT Press,
1960), 2, 3.

35. Lynch, *The Image of the City*, 111.

36. Lynch, *The Image of the City*, 5–6.

37. Lynch, *The Image of the City*, 5, 92.

38. Christopher Alexander, "A City Is Not a Tree," *Architectural Forum*,
April 1965, 58–62.

39. Christopher Alexander, *A Pattern Language: Towns, Buildings,
Construction* (New York: Oxford University Press, 1977), 81.

40. Alexander, *A Pattern Language*, 43.

41. Alexander, *A Pattern Language*, 52.

42. Alexander, *A Pattern Language*, 169.

43. Alexander, *A Pattern Language*, 3.

44. Richard Sennett, *The Fall of Public Man* (New York: Knopf, 1977),
4, 8, 12, 14.

45. Sennett, *The Fall of Public Man*, 15, 16, 301.

46. Sennett, *The Fall of Public Man*, 4, 340.

47. Robert N. Bellah et al., *Habits of the Heart: Individualism and Commitment in American Life* (Berkeley and Los Angeles: University of California Press, 1985), xlii.

48. Bellah et al., *Habits of the Heart*, 43, 44.

49. Bellah et al., *Habits of the Heart*, 177.

50. Bellah et al., *Habits of the Heart*, 50, 127, 113, 139.

51. George L. Kelling and James Q. Wilson, "Broken Windows: The Police and Neighborhood Safety," *The Atlantic*, March 1982, https://www.theatlantic.com/magazine/archive/1982/03/broken-windows/304465.

52. Kelling and Wilson, "Broken Windows."

53. Whyte, *The Social Life of Small Urban Spaces*, 8, 16; William H. Whyte, *City: Rediscovering the Center* (New York: Anchor, 1988), 8, 103.

54. Whyte, *City*, 105, 141, 154.

55. Whyte, *City*, 156.

56. "Project for Public Spaces, Inc.," 1981, p. 1, Rockefeller Brothers Fund Records, box 868, folder 5234 (Project for Public Spaces, The), Rockefeller Archive Center, Sleepy Hollow, NY; "Project for Public Spaces, Inc.: A Prospectus," 1976, p. 1, Downtown-Lower Manhattan Association, Inc., Records, box 162, folder 1554 (Public Spaces Project), Rockefeller Archive Center.

57. "A Proposal to Provide Public Space Improvement Assistance to Neighborhoods in New York City," Project for Public Spaces, Inc., 1979, p. 1, Rockefeller Brothers Fund records, box 868, folder 5234 (Project for Public Spaces, The), Rockefeller Archive Center.

58. E. F. Schumacher, *Small Is Beautiful: A Study of Economics as If People Mattered* (New York: Blond & Briggs, 1973); Amanda Burden, "Greenacre Park: A Study by Project for Public Spaces, Inc.," 1977, pp. 1, 2, Greenacre Foundation Records, Series 2: Projects, box 33, folder 241 (Greenacre Park: A Study by Project for Public Spaces, Inc.), Rockefeller Archive Center.

59. Letter from Marilyn W. Levy to Fred Kent, May 31, 1979, Rockefeller Brothers Fund Records, box 868, folder 5234 (Project for Public Spaces, The), Rockefeller Archive Center; letter from Fred Kent to Marilyn W. Levy, June 11, 1979, Rockefeller Brothers Fund Records, box 868, folder 5234 (Project for Public Spaces, The), Rockefeller Archive Center.

60. "Downtown Management: A New Role for the Private Sector," Project for Public Spaces, Inc., 1983, p. 1, Downtown-Lower Manhattan Association, Inc., Records, box 162, folder 1554 (Public Spaces Project), Rockefeller Archive Center.

61. "Downtown Management: A New Role for the Private Sector," 1, 2.

62. "Public Space Assistance Program," Project for Public Spaces, Inc., February 1980, p. 1, Greenacre Foundation Records, Series 2: Projects, box 33, folder 241 (Greenacre Park: A Study by Project for Public Spaces, Inc.), Rockefeller Archive Center.

63. "Lincoln Memorial Square: Project for Public Spaces, Inc.," July 1976, p. 4, Downtown-Lower Manhattan Association, Inc., Records, box 162, folder 1554 (Public Spaces Project), Rockefeller Archive Center.

64. "Lincoln Memorial Square: Project for Public Spaces, Inc.," 158. We will return to the story of Project for Public Spaces in chapter 3 below.

65. Jay Walljasper, *The Great Neighborhood Book: A Do-It-Yourself Guide to Placemaking* (Gabriola Island, BC: New Society, 2007), 35.

66. Zukin, *The Cultures of Cities*, 31.

67. Zukin, *The Cultures of Cities*, 29, 30, 32.

68. Ray Oldenburg, *The Great Good Place: Cafés, Coffee Shops, Community Centers, General Stores, Bars, Hangouts, and How They Get You through the Day* (New York: Da Capo, 1989), xvi.

69. Oldenburg, *The Great Good Place*, 4, 9.

70. Oldenburg, *The Great Good Place*, 10.

71. Oldenburg, *The Great Good Place*, 20, 22.

72. Oldenburg, *The Great Good Place*, 24, 26, 32.

73. Oldenburg, *The Great Good Place*, 36, 37, 204, 167, 43–44.

74. Oldenburg, *The Great Good Place*, 205, 43, 48, 49, 289, 290.

75. Oldenburg, *The Great Good Place*, 72, 208.

76. Oldenburg, *The Great Good Place*, 214, 60–61.

77. Oldenburg, *The Great Good Place*, 78, 83.

78. Oldenburg, *The Great Good Place*, 293, 284.

79. Oldenburg, *The Great Good Place*, 204–5.

80. Oldenburg, *The Great Good Place*, 264–65, 292.

81. For more on Saul Alinsky's life, see Nicholas von Hoffman, *Radical: A Portrait of Saul Alinsky* (New York: Nation Books, 2010). For a history of the Woodlawn Organization, see John Hall Fish, *Black Power/White Control: The Struggle of the Woodlawn Organization in Chicago* (Princeton, NJ: Princeton University Press, 1973). For more on the "Don't Buy Where You Can't Work" campaigns, see Cheryl Lynn Greenberg, *"Or Does It Explode?" Black Harlem in the Great Depression* (New York: Oxford University Press, 1991), chap. 5; and Brian D. Goldstein, "'The Search for New Form': Black Power and the Making of the Postmodern City," *Journal of American History* 103, no. 2 (September 2016): 375–99, 376. We will reconsider these important histories and their potential importance to alternative understandings of placemaking in chapter 4 below.

82. For the Central Park case, see Sarah Burns, *The Central Park Five: A Chronicle of a City Wilding* (New York: Random House, 2011). For the comparison between the Central Park case and the Kitty Genovese murder, see Buzz Dixon, "Kitty Genovese and the Central Park Five," *Buzz Dixon* (blog), July 5, 2018, https://buzzdixon.com/home/2018/7/5/kitty-genovese-and-the-central-park-five. For the spike in crime in American cities during the late 1980s and early 1990s, see Alfred Blumstein and Joel Wallman, eds., *The Crime Drop in America* (New York: Cambridge University Press, 2006).

83. See, e.g., William Julius Wilson, *When Work Disappears: The World of the New Urban Poor* (New York: Vintage, 1996).

Chapter 2

1. Jennifer M. Gardner, "The 1990–91 Recession: How Bad Was the Labor Market?," *Monthly Labor Review* 117, no. 6 (June 1994): 3–11; Pamela K. Lattimore and Cynthia A. Nahabedian, eds., *The Nature of Homicides: Trends and Changes, Proceedings of the 1996 Meeting of the Homicide Research Working Group*, https://www.ncjrs.gov/pdffiles1/nij/166149.pdf.

2. Kurt Andersen, "The Best Decade Ever? The 1990s, Obviously," *New York Times*, February 6, 2015.

3. "U.S. Underwent a Population Boom in the '90s," *St. Petersburg Times*, April 3, 2001.

4. Edward L. Glaeser and Jesse Shapiro, "Is There a New Urbanism? The Growth of U.S. Cities in the 1990s," Discussion Paper no. 1925 (Cambridge, MA: Harvard Institute of Economic Research, June 2001); "U.S. Underwent a Population Boom in the '90s"; Saskia Sassen, *The Global City: New York, London, Tokyo* (Princeton, NJ: Princeton University Press, 2001).

5. Glaeser and Shapiro, "Is There a New Urbanism?," 2–3, 12.

6. "The Responsive Communitarian Platform," 1991, The Communitarian Network, https://communitariannetwork.org/platform.

7. "The Responsive Communitarian Platform."

8. Quoted in Amitai Etzioni, *The Spirit of Community: The Reinvention of American Society* (New York: Touchstone, 1994), 15.

9. Amitai Etzioni, "Communitarianism," in *The Encyclopedia of Political Thought*, ed. Michael T. Gibbons (New York: Wiley, 2015), https://icps.gwu.edu/sites/icps.gwu.edu/files/downloads/Communitarianism.Etzioni.pdf.

10. Etzioni, "Communitarianism," 119.

11. Etzioni, "Communitarianism," 123, 124.

12. Etzioni, "Communitarianism," 127–28, 129.

13. Etzioni, "Communitarianism," 128.

14. Delores Hayden, *The Power of Place: Urban Landscapes as Public History* (Cambridge, MA: MIT Press, 1995), xii, 7, 9. Hayden was assisted by Sheila Levrant de Bretteville, Donna Graves, Susan King, and Betye Saar on the work chronicled in *The Power of Place*.

15. Hayden, *The Power of Place*, 46, 76, 78.

16. Hayden, *The Power of Place*, 78.

17. Hayden, *The Power of Place*, 246.

18. Zukin, *The Cultures of Cities*, 1, 2.

19. Zukin, *The Cultures of Cities*, 11, 15, 20, 21, 22.

20. "The Blackest Rose Parade Ever? HBCU Marching Bands and Chaka Khan Reign Supreme," *NewsOne*, January 2, 2019, https://newsone.com/playlist/rose-parade-blackest-photos-video.

21. Zukin, *The Cultures of Cities*, 83, 84.

22. Zukin, *The Cultures of Cities*, 265, 280.

23. For accounts of the death of expert-driven urban planning, see Lily M. Hoffman, *The Politics of Knowledge: Activist Movements in Medicine and Planning* (New York: State University of New York Press, 1989); Christopher Klemek, *The Transatlantic Collapse of Urban Renewal: Post Urbanism from New York to Berlin* (Chicago: University of Chicago Press, 2011); and Thomas J. Campanella, "Jane Jacobs and the Death and Life of American Planning," *Places*, April 2011, https://placesjournal.org/article/jane-jacobs-and-the-death-and-life-of-american-planning.

24. William Rich, Kenneth Geiser, Rolf Goetze, and Robert Hollister, "Holding Together: Four Years of Evolution at MIT," *Journal of the American Institute of Planners* 36, no. 4 (July 1970): 242–52, 243.

25. Michael P. Brooks, "Four Critical Junctures in the History of the Urban Planning Profession: An Exercise in Hindsight," *Journal of the American Planning Association* 00, no. 00 (Spring 1988): 241–48; Campanella, "Jane Jacobs and the Death and Life of American Planning."

26. Lynda H. Schneekloth and Robert G. Shibley, *Placemaking: The Art and Practice of Building Communities* (New York: Wiley, 1995), xi, 1.

27. Schneekloth and Shibley, *Placemaking*, 198, 199.

28. Schneekloth and Shibley, *Placemaking*, 2.

29. Schneekloth and Shibley, *Placemaking*, 2, 5.

30. Schneekloth and Shibley, *Placemaking*, 142, 191.

31. Schneekloth and Shibley, *Placemaking*, 192.

32. *The Charter of the New Urbanism*, 1996, Congress for the New Urbanism, https://www.cnu.org/who-we-are/charter-new-urbanism.

33. *The Charter of the New Urbanism*.

34. *The Charter of the New Urbanism*.

35. Andres Duany and Elizabeth Plater-Zyberk, "The Traditional Neighborhood and Urban Sprawl," in *New Urbanism and Beyond: Designing Cities for the Future*, ed. Tigran Haas (New York: Rizzoli, 2008), 64–66, 64.

36. Duany and Plater-Zyberk, "The Traditional Neighborhood and Urban Sprawl," 65.

37. Philip Langdon, "Can Design Make Community?," *The Responsive Community* 7, no. 2 (Spring 1997), 25–37, 26, 28.

38. Langdon, "Can Design Make Community?," 29–30.

39. Langdon, "Can Design Make Community?," 36–37.

40. Author interview with Elizabeth Plater-Zyberk, February 10, 2016.

41. Author interview with Elizabeth Plater-Zyberk, February 10, 2016.

42. James Howard Kunstler, *The Geography of Nowhere: The Rise and Decline of America's Man-Made Landscape* (New York: Simon & Schuster, 1993), 15, 60.

43. Kunstler, *The Geography of Nowhere*, 117, 83.

44. Kunstler, *The Geography of Nowhere*, 250, 251.

45. Kunstler, *The Geography of Nowhere*, 257; "Developing the Cape Fear Economy by Place-Making," May 2011, Cape Fear Economic Development Council, https://capefearedc.org/sponsor-an-event.

46. John O. Norquist, *The Wealth of Cities: Revitalizing the Centers of American Life* (Reading, MA: Addison-Wesley, 1998), 2–3.

47. Norquist, *The Wealth of Cities*, 9, 99, 5.

48. Norquist, *The Wealth of Cities*, 201.

49. Norquist, *The Wealth of Cities*, 190, 191 (quoting Philip Langdon, "The New Neighborly Architecture," *The American Enterprise*, November/December 1996, 42).

50. Tom Daykin, "Riverwalk, Condo Developments Transformed Milwaukee Riverfront," *Milwaukee Journal Sentinel*, June 8, 2014.

51. Tom Daykin, "RiverWalk Turns 20 and Is Still Growing," *Milwaukee Journal Sentinel*, September 17, 2011.

52. Daykin, "RiverWalk, Condo Developments Transformed Milwaukee Riverfront"; Daykin, "RiverWalk Turns 20 and Is Still Growing"; "The Milwaukee Method of Creative Placemaking" (Milwaukee: Greater Milwaukee Committee, Spring 2015), https://9b021b61-8f1f-4c47-8118-1c5db90ecb63.filesusr.com/ugd/7eb190_51899f10148743a3a174236568f3cd52.pdf.

53. "Police Strategy No. 5: Reclaiming the Public Spaces of New York" (New York: New York City Police Department, July 1994), 4.

54. "Police Strategy No. 5," 4, 5.

55. Alex S. Vitale, *City of Disorder: How the Quality of Life Campaign Transformed New York Politics* (New York: New York University Press, 2008), 12.

56. US Commission on Civil Rights, "Police Practices and Civil Rights in New York City" (Washington, DC: US Commission on Civil Rights, August 2000), chap. 5 ("Stop, Question, and Frisk"), https://www.usccr.gov/pubs/nypolice/main.htm.

57. Giuliani quoted in Vitale, *City of Disorder*, 29; Timothy Williams and Joseph Goldstein, "In Baltimore Report, Justice Dept. Revives Doubt about Zero-Tolerance Policing," *New York Times*, August 10, 2016, http://www.nytimes.com/2016/08/11/us/baltimore-police-zero-tolerance-justice-department.html?_r=0.

58. Robert Putnam, "The Prosperous Community: Social Capital and Public Life" (1993), *The American Prospect*, December 19, 2001, https://prospect.org/infrastructure/prosperous-community-social-capital-public-life.

59. Putnam, "The Prosperous Community."

60. Putnam, "The Prosperous Community."

61. Putnam, "The Prosperous Community."

62. Putnam, "The Prosperous Community."

63. Putnam, "The Prosperous Community."

64. Francis Fukuyama, *Trust: The Social Virtues and the Creation of Prosperity* (New York: Free Press, 1995), xiii.

65. Fukuyama, *Trust*, 4.

66. Fukuyama, *Trust*, 4, 5.

67. Fukuyama, *Trust*, 6.

68. Fukuyama, *Trust*, 11, 33.

69. Fukuyama, *Trust*, 307, 362.

70. Robert D. Putnam, *Bowling Alone: The Collapse and Revival of American Community* (New York: Simon & Schuster, 2000), 19.

71. Putnam, *Bowling Alone*, 308.

72. Putnam, *Bowling Alone*, 308 (including Jacobs quote).

73. Putnam, *Bowling Alone*, 23, 24.

74. Putnam, *Bowling Alone*, 312, 319, 322.

75. Putnam, *Bowling Alone*, 411.

76. Putnam, *Bowling Alone*, 403, 412.

77. After School Matters, https://www.afterschoolmatters.org/about-us/general.

78. William Julius Wilson, *The Truly Disadvantaged: The Inner City, the Underclass, and Public Policy* (Chicago: University of Chicago Press, 1987), 60; Wilson, *When Work Disappears*, xiii, 54.

79. "N30 Black Bloc Communiqué," December 4, 1999, ACME Collective, https://theanarchistlibrary.org/library/acme-collective-n30-black-bloc-communique.

80. Renato Ruggiero, "Reflections from Seattle," in *The WTO After Seattle*, ed. Jeffrey J. Schott (Washington, DC: Institute for International Economics, 2000): xiii–xvii, xiv, xv.

81. "Programs," n.d., After School Matters, https://www.afterschoolmatters.org/teens/programs.

Chapter 3

1. "Bowling in Groups," *Publisher's Weekly*, August 14, 2000, 273.

2. Ray Oldenburg, *Celebrating the Third Place: Inspiring Stories about the "Great Good Places" at the Heart of Our Communities* (New York: Marlowe, 2001), 2.

3. Oldenburg, *Celebrating the Third Place*, 5.

4. Oldenburg, *Celebrating the Third Place*, 2.

5. Holcepl quoted in Oldenburg, *Celebrating the Third Place*, 124.

6. Oldenburg, *Celebrating the Third Place*, 125.

7. Richard Florida, *The Flight of the Creative Class: The New Global Competition for Talent* (New York: HarperBusiness, 2005), 41.

8. Florida, *The Rise of the Creative Class*, 4, 5, 44.

9. Florida, *The Rise of the Creative Class*, 9, 42.

10. Florida, *The Rise of the Creative Class*, 6, 7, 12.

11. Florida, *The Rise of the Creative Class*, 12, 173, 185, 225, 231.

12. Florida, *The Rise of the Creative Class*, 81, 82, 74, 76.

13. Richard Florida, *Cities and the Creative Class* (New York: Routledge, 2004), 117. The year 2005 saw the publication of Florida's *The Flight of the Creative Class*.

14. Richard Florida, *Who's Your City? How the Creative Economy Is Making Where to Live the Most Important Decision of Your Life* (New York: Basic, 2008), 43, 66.

15. Florida, *Who's Your City?*, 159.

16. Florida, *Cities and the Creative Class*, 30.

17. Florida, *Cities and the Creative Class*, 125.

18. Richard Florida, *The Great Reset: How New Ways of Living and Working Drive Post-Crash Posterity* (New York: HarperBusiness, 2010), 5.

19. Florida, *The Great Reset*, 7.

20. Florida, *The Great Reset*, 105, 6, 74, 76.

21. Florida, *The Great Reset*, 46.

22. Florida, *The Great Reset*, 181.

23. Florida, *The Great Reset*, 155, 119, 181.

24. Cynthia Nikitin, "All Placemaking Is Creative: How a Shared Focus on Space Builds Vibrant Communities," March 3, 2013, Project for Public Spaces, https://www.pps.org/article/placemaking-as-community-creativity-how-a-shared-focus-on-place-builds-vibrant-destinations.

25. *How to Turn a Place Around: A Handbook for Creating Successful Public Spaces* (New York: Project for Public Spaces, 2000), 13.

26. *How to Turn a Place Around*, 16, 17.

27. *How to Turn a Place Around*, 20, 47, 71.

28. *How to Turn a Place Around*, 47, 49.

29. *How to Turn a Place Around*, 30, 31, 55, 63.

30. *How to Turn a Place Around*, 67.

31. "Public Markets as a Vehicle for Social Integration and Upward Mobility: Phase I Report: An Overview of Existing Programs and Assessment of Opportunities" (New York: Project for Public Spaces; Washington, DC: Partners for Livable Communities, September 2003), 5, https://s3.amazonaws.com/aws-website-ppsimages-na05y/pdf/Ford_Report.pdf.

32. "Public Markets as a Vehicle for Social Integration and Upward Mobility," 5.

33. "Public Markets as a Vehicle for Social Integration and Upward Mobility," 5, 7.

34. "Public Markets as a Vehicle for Social Integration and Upward Mobility," 27.

35. "Mayors and Businesses Driving Economic Growth," Business Council Best Practice Report (Washington, DC: US Conference of Mayors, 2015), 27, http://www.muniservices.com/wp-content/uploads/US_Mayors_and-Businesses_Driving_Economic_Growth.pdf.

36. Walljasper, *The Great Neighborhood Book*, 1.

37. Robert E. Park, Ernest W. Burgess, and Roderick D. McKenzie, *The City: Suggestions for Investigation of Human Behavior in the Urban Environment* (1925; reprint, Chicago: University of Chicago Press, 1967).

38. Max Weber, *Economy and Society: An Outline of Interpretive Sociology*, ed. Guenther Roth and Claus Wittich (1921; reprint, Berkeley and Los Angeles: University of California Press, 1978); Émile Durkheim, *The Division of Labor in Society* (1893; reprint, New York: Free Press, 1997); Georg Simmel, "The Metropolis and Mental Life" (1903), in *Metropolis: Center and Symbol of Our Times*, ed. Philip Kasinitz (New York: New York University Press, 1995); C. B. Fawcett, "British Conurbations in 1921," *Sociological Review* 14 (1922): 111–22.

39. Park, Burgess, and McKenzie, *The City*, 13.

40. Park, Burgess, and McKenzie, *The City*.

41. *The City: Suggestions for Investigation of Human Behavior in the Urban Environment*; Harvey Zorbaugh, *The Gold Coast and the Slum* (Chicago: University of Chicago Press, 1929); Paul G. Cressey, *The Taxi-Dance Hall* (Chicago: University of Chicago Press, 1932); Walter Reckless, *Vice in Chicago* (Chicago: University of Chicago Press, 1933).

42. Park, Burgess, and McKenzie, *The City*, 190.

43. Park, Burgess, and McKenzie, *The City*, 154.

44. Morris Janowitz, *The Community Press in an Urban Setting: The Social Elements of Urbanism* (1952), 2nd ed. (Chicago: University of Chicago Press, 1967); Gerald D. Suttles, *The Social Order of the Slum: Ethnicity and Territory in the Inner City* (Chicago: University of Chicago Press, 1968).

45. More contemporary examples examining the role of local institutions as important sites of community interaction and constitution include Robert Smith, *Mexican New York: Transnational Lives of New Immigrants* (Berkeley and Los Angeles: University of California Press, 2005); Nicole Marwell, *Bargaining for Brooklyn: Community Organizations in the Entrepreneurial City* (Chicago: University of Chicago

Press, 2007); Mary E. Pattillo, *Black on the Block: The Politics of Race and Class in the City* (Chicago: University of Chicago Press, 2007); Mario Small, *Unanticipated Gains: Origins of Network Inequality in Everyday Life* (New York: Oxford University Press, 2009); and Cid Gregory Martinez, *The Neighborhood Has Its Own Rules: Latinos and African Americans in South Los Angeles* (New York: New York University Press, 2016).

46. See, e.g., Terry Nichols Clark, ed., *Urban Innovation: Creative Strategies for Turbulent Times* (Thousand Oaks, CA: Sage, 1994); Terry Nichols Clark and Vincent Hoffmann-Martinot, eds., *The New Political Culture* (Boulder, CO: Westview, 1998); Terry Nichols Clark, Richard Lloyd, Kenneth K. Wong, and Pushpam Jain, "Amenities Drive Urban Growth," *Journal of Urban Affairs* 24 (2002): 493–515; Terry Nichols Clark, ed., *The City as an Entertainment Machine*. (Lanham, MD: Lexington, 2011); and Daniel Aaron Silver and Terry Nichols Clark, *Scenescapes: How Qualities of Place Shape Social Life* (Chicago: University of Chicago Press, 2016).

47. Silver and Clark, *Scenescapes*.

48. Richard Lloyd, *Neo-Bohemia: Art and Commerce in the Post-Industrial City* (New York: Routledge, 2006), 56.

49. Richard Florida and Gary Gates, "Technology and Tolerance: The Importance of Diversity to High Technology Growth," in Clark, ed., *The City as an Entertainment Machine*, 157–78.

50. Although he now teaches at Harvard, Sampson was at the University of Chicago from 1991 to 2002, overlapping with Clark and Lloyd.

51. Often cited is Robert J. Sampson, "Family Management and Child Development: Insights from Social Disorganization Theory," in *Facts, Framework, and Forecasts: Advances in Criminological Theory* (vol. 3), ed. Joan McCord (New Brunswick, NJ: Transaction, 1992), 62–93.

52. Robert J. Sampson, *Great American City: Chicago and the Enduring Neighborhood Effect* (Chicago: University of Chicago Press, 2012), 5.

53. Sampson, *Great American City*, 426, 421.

54. Sampson, *Great American City*, 23, 179.

55. "Social Impact of the Arts Project," n.d., Scholarly Commons, University of Pennsylvania, https://repository.upenn.edu/siap.

56. Carl Bialik, "Philadelphia Is Bouncing Back from Problems Still Plaguing Cleveland," *FiveThirtyEight*, July 26, 2016, http://fivethirtyeight.com/features/philadelphia-is-bouncing-back-from-problems-still-plaguing-cleveland; Federal Reserve Bank of Philadelphia, Research Department, "The Industrial Evolution: Two Decades of Change in the Philadelphia Metro Area's Economy" (Philadelphia: Federal Reserve Bank of Philadelphia, Research Department, January 2002), 6.

57. Mark J. Stern and Susan C. Seifert, "Community Revitalization and the Arts in Philadelphia," Working paper, 1998, 4, Scholarly Commons, University of Pennsylvania, https://repository.upenn.edu/siap_culture_builds_community/8.

58. Stern and Seifert, "Community Revitalization and the Arts in Philadelphia," 5, 6, 7.

59. Stern and Seifert, "Community Revitalization and the Arts in Philadelphia," 8.

60. Stern and Seifert, "Community Revitalization and the Arts in Philadelphia," 7.

61. Mark J. Stern and Susan C. Seifert, "Cultural Participation and Communities: The Role of Individual and Neighborhood Effects," Conference paper, 2000, 1, Scholarly Commons, University of Pennsylvania, https://repository.upenn.edu/siap_culture_builds_community/12.

62. Stern and Seifert, "Cultural Participation and Communities," 1, 2, 4.

63. Mark J. Stern and Susan C. Seifert, "The Dynamics of Cultural Participation: Metropolitan Philadelphia, 1996–2004," Working paper, October 2005, 27, Scholarly Commons, University of Pennsylvania, https://repository.upenn.edu/siap_dynamics/1.

64. Mark J. Stern and Susan C. Seifert, "Cultivating 'Natural' Cultural Districts," Policy brief, 2007, 1, Scholarly Commons, University of Pennsylvania, https://repository.upenn.edu/siap_revitalization/4.

65. Stern and Seifert, "Cultivating 'Natural' Cultural Districts," 1.

66. Stern and Seifert, "Cultivating 'Natural' Cultural Districts," 1.

67. Jeremy Nowak, "Creativity and Neighborhood Development: Strategies for Community Investment," Report, 2007, 1, Scholarly Commons, University of Pennsylvania, https://repository.upenn.edu/siap_revitalization/2.

68. Nowak, "Creativity and Neighborhood Development," 1, 2, 6.

69. Nowak, "Creativity and Neighborhood Development," 4.

70. Landesman quoted in Jane Chu and Jason Schupbach, "Our Town: Supporting the Arts in Communities throughout the United States," *Community Development Investment Review*, December 2014, 65–71, 66.

71. Diana Jean Schemo, "Endowment Ends Program Helping Individual Artists," *New York Times*, November 3, 1994; Mark Bauerlein and Ellen Grantham, eds., *National Endowment of the Arts: A History, 1965–2008* (Washington, DC: National Endowment for the Arts, 2009), 111.

72. Ann Markusen, "Creative Cities: A Ten-Year Research Agenda," *Journal of Urban Affairs* 36, no. 2 (2014): 1–23.

73. Ann McQueen, "An Interview with Rocco Landesman," Winter 2013, Grantmakers in the Arts, https://www.giarts.org/article/interview-rocco-landesman.

74. "A Conversation with Ann Markusen: Her Word on Creative Placemaking," *Carolina Planning Journal* 41 (2016): 18–21, 19.

75. It should be noted that the term *creative placemaking* was not

something the authors came up with. Referring to the NEA, Markusen has noted that "creative placemaking was their rubric." Author interview with Ann Markusen and Anne Gadwa Nicodemus, December 5, 2016.

76. Markusen and Gadwa, "Creative Placemaking," 3.

77. Markusen and Gadwa, "Creative Placemaking," 3, 7.

78. Markusen and Gadwa, "Creative Placemaking," 3.

79. Markusen and Gadwa, "Creative Placemaking," 3, 5, 8.

80. Markusen and Gadwa, "Creative Placemaking," 3, 10, 15.

81. Markusen and Gadwa, "Creative Placemaking," 13, 14.

82. Markusen and Gadwa, "Creative Placemaking," 10, 14.

83. Markusen and Gadwa, "Creative Placemaking," 29.

84. "New Funding and Resources for Creative Placemaking," Press release, July 14, 2015, National Endowment for the Arts, https://www.arts.gov/news/2015/new-funding-and-resources-creative-placemaking; author interview with Ann Markusen and Anne Gadwa Nicodemus, December 5, 2016.

85. Jason Schupbach, "Creative Placemaking," *Economic Development Journal* 14, no. 4 (Fall 2015): 28–33, 29 (quoting Chu).

86. Rip Rapson, "Arts and Culture in Detroit: Central to Our Past and Our Future," *Community Development Investment Review*, December 2014, 73–75, 75.

87. "Creative Placemaking: An Interview with the Ford Foundation," *Community Development Investment Review*, December 2014, 11–16, 11, 13.

88. "Creative Placemaking: An Interview with the Ford Foundation," 11, 13.

89. Author interview with Susan Silberberg, October 26, 2016.

90. Silberberg, "Places in the Making."

91. Silberberg, "Places in the Making," 1, 2, 3, 7.

92. Silberberg, "Places in the Making," 10, 11.

93. Silberberg, "Places in the Making," 28, 29, 32, 33, 34.

94. "Partners," n.d., ArtPlace America, https://www.artplaceamerica.org/about/partners; Danya Sherman, "Can Taxpayer-Funded Placemaking Survive Trump?," January 23, 2017, Next City, https://nextcity.org/daily/entry/nea-creative-placemaking-our-town-grants-trump.

95. Jamie Bennett, "Creative Placemaking in Community Planning and Development: An Introduction to ArtPlace America," *Community Development Investment Review*, December 2014, 77–84, 77, 78.

96. Bennett, "Creative Placemaking in Community Planning and Development," 78, 80.

97. Pamela Espeland, "FLOW Northside Arts Crawl Enters Second Decade; Nimbus Has New Digs," MinnPost, July 27, 2016, https://www.minnpost.com/artscape/2016/07/flow-northside-arts-crawl-enters-second-decade-nimbus-has-new-digs; "West Broadway Business and Area Coalition Receives 2015 ArtPlace America Grant," Press release, July 13, 2015, West Broadway Business and Area Coalition, http://westbroadway.org/artplace.

98. Eric Roper, "New Public Square to Liven Up W. Broadway in Minneapolis," *Star Tribune*, July 30, 2016, http://www.startribune.com/new-public-square-to-liven-up-w-broadway-in-minneapolis/388725601.

99. Rip Rapson, "Creative Placemaking: The Next Phase for ArtPlace," Address to the ArtPlace America Summit on Creative Placemaking, Los Angeles, CA, March 4, 2014, http://kresge.org/sites/default/files/library/creative-placemaking-the-next-phase-for-artplace.pdf.

100. Roger Bybee, "Occupy the Hood: Fighting for Those 'at the Bottom of the Bottom,'" *In These Times*, December 1, 2011, http://inthesetimes.com/working/entry/12355/occupy_the_hood_fighting_for_those_atht_the_bottom_of_the_bottom.

Chapter 4

1. Richard Florida, *The New Urban Crisis: How Our Cities Are Increasing Inequality, Deepening Segregation, and Failing the Middle Class—and What We Can Do about It* (New York: Basic, 2017), xvi, xx.

2. Florida, *The New Urban Crisis*, 189.

3. Florida, *The New Urban Crisis*, 190, 194.

4. Thomas Frank, "Dead End on Shakin' Street," *The Baffler*, no. 20 (July 2012), https://thebaffler.com/salvos/dead-end-on-shakin-street.

5. Ian David Moss, "Creative Placemaking Has an Outcomes Problem," May 9, 2012, Createquity, http://createquity.com/2012/05/creative-placemaking-has-an-outcomes-problem.

6. Moss, "Creative Placemaking Has an Outcomes Problem."

7. Matt Tyrnauer, dir., *Citizen Jane: Battle for the City* (2016).

8. Brian Tochterman, "Theorizing Neoliberal Urban Development: A Genealogy from Richard Florida to Jane Jacobs," *Radical History Review*, no. 112 (Winter 2012): 65–87, 73; Max Page, "Introduction: More Than Meets the Eye," in *Reconsidering Jane Jacobs*, ed. Max Page and Timothy Mennel (New York: Routledge, 2017), 3–14, 8.

9. Shelleen Greene, "Art City Asks: Fidel Verdin," *Milwaukee Journal-Sentinel*, September 15, 2016, https://www.jsonline.com/story/entertainment/arts/art-city/2016/09/15/art-city-asks-fidel-verdin-true-skool/90353492.

10. Author interview with Fidel Verdin, August 2, 2017.

11. Author interview with Fidel Verdin, August 2, 2017.

12. Author interview with Fidel Verdin, August 2, 2017.

13. Author interview with Fidel Verdin, August 2, 2017.

14. Author interview with Fidel Verdin, August 2, 2017; Author interview with Fidel Verdin, March 13, 2019.

15. Author interview with Fidel Verdin, March 13, 2019.

16. Author interview with Fidel Verdin, March 13, 2019. For the work of Venice Williams, see Annysa Johnson, "Herb Farm Is Latest Project from Milwaukee Visionary Venice Williams," *Milwaukee Journal-Sentinel*, December 17, 2013, http://archive.jsonline.com/news/milwaukee/herb-farm-is-latest-project-from-milwaukee-visionary-venice-williams-b99156589z1-236139121.html.

17. Author interview with Fidel Verdin, March 13, 2019; "Urban Agriculture Code Audit: Milwaukee, Wisconsin," April 2012, US Environmental Protection Agency, https://city.milwaukee.gov/ImageLibrary/Groups/cityDCD/Urban-Agriculture/pdfs/MilwaukeeCodeAudit_acknowledge.pdf.

18. Monica M. White, *Freedom Farmers: Agricultural Resistance and the Black Freedom Movement* (Chapel Hill: University of North Carolina Press, 2019), 5, 118.

19. Joseph B. Walzer, "Brewing," n.d., *Encyclopedia of Milwaukee*, https://emke.uwm.edu/entry/brewing.

20. John Gurda, "Eat, Drink, and Be Prosperous: A Short History of the Food and Beverage Industry in the Milwaukee Region—the Seven Counties of Southeastern Wisconsin," December 7, 2010, Report commissioned by the Milwaukee 7, 2, http://c.ymcdn.com/sites/www.fabmilwaukee.com/resource/resmgr/docs/john_gurda_f&b_history_2010.pdf (link inactive).

21. Edward S. Kerstein, *My South Side* (Milwaukee: Milwaukee Journal, 1976), 2.

22. Marketing and Facilities Research Branch, US Department of Agriculture, in cooperation with the Department of Agricultural Economics, University of Wisconsin, "The Wholesale Produce Market at Milwaukee, Wis." (Milwaukee: Legislative Reference Bureau, Milwaukee City Hall, January 1950), 1, 2, 3, 6, 7.

23. Agricultural Experiment Station, University of Wisconsin, Madison, "Milwaukee Wholesale Fruit & Vegetable Market" (Milwaukee: Legislative Reference Bureau, Milwaukee City Hall, February 1948), 1.

24. "Market Stalls Open for 1940," *Milwaukee Journal*, June 16, 1940.

25. "Market Stalls Open for 1940."

26. "A Report on Milwaukee Municipal Farmers' Markets" (Milwaukee: Legislative Reference Bureau, Milwaukee City Hall, 1973), 1, 2.

27. Marilyn Gardner, "The Center St. Crowd," *Milwaukee Journal*, June 25, 1980.

28. "Farm Mart to Feature Amenities," *Milwaukee Journal*, December 4, 1979; "Fondy Farm Market Heralds Area Rebirth," *Milwaukee Sentinel*, June 23, 1981.

29. "Fondy Farmers' Market" (advertisement), Milwaukee Sentinel, October 8, 1988.

30. John Gurda, *The Making of Milwaukee* (Milwaukee: Milwaukee County Historical Society, 1999), 314.

31. "War Was Seed for Gardens," *Milwaukee Journal*, March 23, 1978.

32. Steven D. Brachman, Text of Presentation Given before the Policy Committee, January 30, 1979, Henry W. Maier Papers, box 166, folder 1, Special Collections, University of Wisconsin, Milwaukee.

32. Minutes from Milwaukee Urban League Meeting, November 13, 1962, Milwaukee Urban League Records, box 12, folder 45, Special Collections, University of Wisconsin, Milwaukee.

34. Memorandum from Joseph E. Downey to Carl Smith, January 17, 1963, Milwaukee Urban League Records, box 12, folder 45, Special Collections, University of Wisconsin, Milwaukee.

35. Letter from Carl S. Smith to Urban League 4-H Leaders, March 23, 1963, Milwaukee Urban League Records, box 12, folder 45, Special Collections, University of Wisconsin-Milwaukee.

36. "Recommendation concerning Organization of Adult Education and Extension Activities," September 3, 1963, UW-Extension Records, 1896–, box 11, folder 20, Special Collections, University of Wisconsin, Milwaukee.

37. "A Proposal for the Experiment in Continuing Education for Selected Citizens in the 'Northtown A' Area," Urban Work Group, Extension-Milwaukee, UW-Extension Records, box 14, folder 11, Special Collections, University of Wisconsin, Milwaukee.

38. UWEX Urban Committee, "Poverty Proposals," June 9, 1964, UW-Extension Records, box 14, folder 11, Special Collections, University of Wisconsin, Milwaukee.

39. "1974 Milwaukee County Family Garden Project Progress Report," Henry Reuss Papers, box 16, folder 14; City of Milwaukee Interdepartmental Correspondence, Bridget Bannon to File, RE: City Garden Plot Program, January 16, 1978, Henry W. Maier Papers, box 86, folder 30, Special Collections, University of Wisconsin, Milwaukee.

40. Memorandum from Michael J. Brady to Park West Task Force, March 1, 1977, Henry Reuss Papers, box 16, folder 14, Special Collections, University of Wisconsin, Milwaukee.

41. Letter from Earnestine Black to Mayor Maier, n.d., Henry W. Maier Papers, box 86, folder 30, Special Collections, University of Wisconsin, Milwaukee; Letter from Robert W. Skiera to Earnestine Black, March 14, 1978, Henry W. Maier Papers, box 86, folder 30, Special Collections, University of Wisconsin, Milwaukee.

42. "How to Save Money by Growing Your Own Food," Shoots 'n Roots informational brochure, Henry W. Maier Papers, box 86, folder 30, Special Collections, University of Wisconsin-Milwaukee.

43. US Department of Agriculture, Press Release, n.d., Henry W. Maier Papers, box 86, folder 30, Special Collections, University of Wisconsin, Milwaukee; "Congress Passes a Bill to Aid Urban Gardens," *New York Times*, July 7, 1976; "Get a Garden and Save," *Sarasota Herald-Tribune*, April 27, 1978.

44. Letter from Steven D. Brachman to Mayor Henry Maier, April 13, 1978, Henry W. Maier Papers, box 86, folder 30, Special Collections, University of Wisconsin, Milwaukee.

45. Steven D. Brachman, "FY Plan of Work for the Urban Garden Program, Cooperative Extension Programs, Milwaukee, WI," 2, copy in authors' possession.

46. Brachman, "FY Plan of Work for the Urban Garden Program," 3, 4.

47. Brachman, Text of Presentation Given before the Policy Committee.

48. "City Spent $776,716 in '78 to Maintain Its Vacant Lots," *Milwaukee Journal*, January 24, 1979.

49. Brachman, Text of Presentation Given before the Policy Committee.

50. "Home Gardening Growing Rapidly," *Milwaukee Journal*, December 9, 1979.

51. Letter from Steven D. Brachman to Leila Fraser, June 20, 1979, Henry W. Maier Papers, box 166, folder 1, Special Collections, University of Wisconsin, Milwaukee.

52. "Field Day: Bus Makes Circuit of Garden Tracks," *Milwaukee Journal*, July 31, 1979.

53. "4-H Youths' Gardens Produce Profits, Too," *Milwaukee Journal*, August 5, 1981.

54. Steve Brachman, "Developing Our Community through Gardening in Milwaukee," *Journal of Community Gardening* 1, no. 4 (December 1982): 4, 5.

55. Remarks by Mayor Henry W. Maier, Shoots 'n Roots Community Gardening Groundbreaking Program, May 5, 1983, Henry W. Maier Papers, box 176, folder 11.

56. Brilton Rodrieguez, "Soil Condition and the Productivity of Two Community Garden Plots in the City of Milwaukee, Wisconsin during the 1986 Growing Season" (MA thesis, University of Wisconsin, Milwaukee, May 1989), 16.

57. "Lend a Trowel," *Milwaukee Journal*, May 24, 1992.

58. "Back to Basics," *Milwaukee Journal*, June 17, 1994.

59. Elizabeth Royte, "Street Farmer," *New York Times Magazine*, July 1, 2009, https://www.nytimes.com/2009/07/05/magazine/05allen-t.html; "Chicago Public Schools Celebrate Farm to School Month," October 20, 2015, Healthy Schools Campaign, https://healthyschoolscampaign.org/chicago-school-food/chicago-public-schools-celebrates-farm-to-school-month.

69. Author interview with Will Allen, November 1, 2013.

61. Growing Power, Fact sheet, n.d., copy in authors' possession; Jen Hunholz, "Global Giving," *M Magazine*, April 2015, 69.

62. Author interview with Will Allen, November 1, 2013.

63. Author interview with Will Allen, November 1, 2013.

64. Author interview with Will Allen, November 1, 2013.

65. Author interview with Will Allen, November 1, 2013.

66. Author interview with Will Allen, November 1, 2013.

67. Will Allen with Charles Wilson, *The Good Food Revolution: Growing Healthy People, Food, and Communities* (New York: Gotham, 2012), 16; "Council Forges Ahead on Directives to Address Black Male Unemployment," February 29, 2012, News Release, City of Milwaukee, https://city.milwaukee.gov/ImageLibrary/Groups/ccCouncil/2012PDF/UnemploymentTaskForcerel.pdf.

68. Correspondence with Sharon Adams, January 31, 2018.

69. "Community Gardens Help Raise Food and Spirits," *Walnut Observer*, March 2001, 3, Walnut Way Conservation Corp. Records, box 1, folder 8, Special Collections, University of Wisconsin, Milwaukee.

70. "Growing Power/Opportunity in Walnut Way," *Walnut Observer*, May 2001, 3, Walnut Way Conservation Corp. Records, box 1, folder 8, Special Collections, University of Wisconsin, Milwaukee.

71. "The Walnut Way Youth Garden Project," July 2002, Walnut Way Conservation Corp. Records, box 1, folder 4, Special Collections, University of Wisconsin, Milwaukee.

72. "Walnut Way Conservation Corp.," n.d., Brochure, Walnut Way Conservation Corp. Records, box 1, folder 2, Special Collections, University of Wisconsin, Milwaukee.

73. "Spring Tulip Sale: From Our Garden to You," May 2003, Walnut Way Conservation Corp. Records, box 1, folder 2, Special Collections, University of Wisconsin, Milwaukee.

74. "Walnut Way Market," *Observer News Flyer*, Summer 2004, 1, Walnut Way Conservation Corp. Records, box 1, folder 8, Special Collections, University of Wisconsin, Milwaukee.

75. "Vacant Lot to Bear Fruit in Walnut Way," *Neighborhood News*, Summer 2006, Walnut Way Conservation Corp. Records, box 1, folder 8, Special Collections, University of Wisconsin, Milwaukee.

76. "Walnut Way Gardens to Market Program," *Neighborhood News*, Autumn 2007, Walnut Way Conservation Corp. Records, box 1, folder 8, Special Collections, University of Wisconsin, Milwaukee.

77. "The Walnut Way Story Project: Caring Neighbors Make Good Communities," 2010, 2, Walnut Way Conservation Corp. Records, box 1, folder 3, Special Collections, University of Wisconsin, Milwaukee.

78. Michael Holloway, "New Innovation & Wellness Commons Opens," *Urban Milwaukee*, November 3, 2015, https://urbanmilwaukee.com/2015/11/03/new-innovation-wellness-commons-opens.

79. Jackson Dufault, Conor Stanley, and Khaleeq Sattaar El, "Walnut Way Executive Director Promotes Economic Development, Community," *Milwaukee Neighborhood News Service*, May 14, 2018, https://milwaukeenns.org/2018/05/14/walnut-way-executive-director-promotes-economic-development-community; Lauren Anderson, "Walnut Way to begin work on second phase of Innovation and Wellness Commons," *Milwaukee BizTimes*, November 22, 2019, https://biz times.com/walnut-way-to-begin-work-on-second-phase-of-innova tions-and-wellness-commons.

80. Author interview with James Godsil, May 10, 2010.

81. Author interview with James Godsil, May 10, 2010.

82. Tom Daykin, "Some Souring on Sweet Water's City Financing Deal," *Milwaukee Journal-Sentinel*, April 7, 2012, http://archive.jsonline .com/business/some-souring-on-sweet-waters-city-financing-deal -f64rugb-146527625.html; Stacy Vogel Davis, "Sweet Water Organics Stops Operation, Defaults on City Loan," *Milwaukee Business Journal*, May 10, 2013, https://www.bizjournals.com/milwaukee/news /2013/05/10/sweet-water-organics-stops-operation.html; Tom Daykin, "Bay View's Unique Vibe in for a Change," *Milwaukee Journal-Sentinel*, March 1, 2017, https://www.jsonline.com/story/money/real -estate/commercial/2017/03/01/some-welcome-high-end-apart ments-bay-view-others-wince/98205082.

83. "Radical (Re)Construction of Community: Food, Art, and Regenerative Placemaking on Chicago's South Side," March 4, 2018, Sweet Water Foundation, https://www.sweetwaterfoundation.com/latest -news-1/2017/12/12/zyw44vtwgfiq8s5ejdck2j0pod5f55 (link inactive).

84. Janette Kim, "Harvesting Change," *ARPA Journal*, May 24, 2018, http://www.arpajournal.net/harvesting-change.

85. Yakini quoted in "Growing Power Awarded $5 Million Grant to Grow Community Food Projects across U.S.," Press release, September 8, 2012, W. K. Kellogg Foundation, https://www.wkkf.org/news -and-media/article/2012/09/growing-power-awarded-$5-million -grant.

86. "Growing Power Awarded $5 Million Grant"; Laura J. Lawson, *City Bountiful: A Century of Community Gardening in America* (Berkeley and Los Angeles: University of California Press, 2005), 23–26.

87. Yakini quoted in Page Pfleger, "Detroit's Urban Farms: Engines of Growth, Omens of Change," January 11, 2018, WHYY, https://whyy.org /segments/detroits-urban-farms-engines-growth-omens-change.

88. Brian Allnutt, "A Black-Led Food Co-Op Grows in Detroit," January 21, 2019, CityLab, https://www.citylab.com/equity/2019/01/black -owned-food-coop-detroit-dbcfsn-d-town-farm-yakini/580819.

89. Yakini quoted in Melinda Clynes, "'A Co-Op for the People: The Rocky Process of Developing the Detroit People's Co-Op,' Model D," n.d., Michigan Good Food Fund, http://migoodfoodfund.org/news -items/a-co-op-for-the-people-the-rocky-process-of-developing-the -detroit-peoples-food-co-op-model-d.

90. Correspondence with Johanna Gilligan, March 8, 2018; "Leadership Programs," Grow Dat Youth Farm, https://growdatyouthfarm. org/our-program; author interview with Jabari Brown, November 16, 2016.

91. Author interview with Jabari Brown, November 16, 2016; Correspondence with Johanna Gilligan, March 8, 2018; Jason Bethea, "Grow Dat Organizational History," Video, https://vimeo.com/85393110.

92. Correspondence with Johanna Gilligan, March 8, 2018; Ian McNulty, "Hollygrove Market Is Closing; Future of the New Orleans Hub for Urban Farming Uncertain," *New Orleans Advocate*, February 26, 2018, https://www.theadvocate.com/new_orleans/entertainment _life/food_restaurants/article_c915ed70-1b31-11e8-be7f-c74f98276 34e.html.

93. "Cleveland Neighborhoods by the Numbers," August 2016, Western Reserve Land Conservancy, https://www.wrlandconservancy.org /wp-content/uploads/2016/08/ClevelandPropertyInventory_issuu _updated122016.pdf.

94. David Sax, "How a Winery Is Transforming Cleveland's Most Notorious Neighborhood," *Good*, February 12, 2015, https://www.good.is /features/chateau-hough-cleveland-winery.

95. Jeremy Nobile, "Ohio Winemakers Toast Industry's Booming Biz," *Crain's Cleveland Business*, October 15, 2017, https://www.crains cleveland.com/article/20171015/news/138781/ohio-winemakers -toast-industry%E2%80%99s-booming-biz.

96. Adele Peters, "Turning a Vacant Cleveland House into a Fancy Farm," *Fast Company*, November 12, 2013, https://www.fastcompany .com/3021341/turning-a-vacant-cleveland-house-into-a-fancy-farm.

97. Billy Hallal, "Innovations from Cleveland's Urban Farms Are Taking Root around the World," *FreshWater*, November 9, 2017, http://www .freshwatercleveland.com/features/urbanfarm110917.aspx.

98. David Gutierrez, "Appetite for Change Brings Fresh Ideas to the Table," *Minneapolis Star Tribune*, September 4, 2018, http://www .startribune.com/appetite-for-change-brings-fresh-ideas-to-the -table/492202541.

99. Appetite for Change, "Grow Food" lyrics, Genius, https://genius .com/Appetite-for-change-grow-food-lyrics.

100. Appetite for Change, "Grow Food" lyrics.

Chapter 5

1. "The Rooftop Farm Ecosystem," n.d., Eagle Street Rooftop Farm, https://rooftopfarms.org/greenroof-ecosystem.

2. Anastasia Cole Plakias, *The Farm on the Roof: What Brooklyn*

Grange Taught Us about Entrepreneurship, Community, and Growing a Sustainable Business (New York: Avery, 2016), 3, 4.

3. Plakias, *The Farm on the Roof*, 261, 263.

4. Plakias, *The Farm on the Roof*, 264, 265.

5. "DEP Awards $3.8 Million in Grants for Community-Based Green Infrastructure Program Projects," Press release, June 9, 2011, New York City Department of Environmental Protection, http://www.nyc.gov /html/dep/html/press_releases/11-46pr.shtml.

6. "Urban Agriculture," n.d., City Colleges of Chicago, https://www .ccc.edu/menu/Pages/Green-Space.aspx; Megy Karydes, "Chicago's Growing Home Is Using Food and Farming to Help Nourish More Than the Body," *Forbes*, April 8, 2019, https://www.forbes.com/sites/ megykarydes/2019/04/08/chicagos-growing-home-is-using-food -and-farming-to-help-nourish-more-than-the-body/#ca50bc06b90 8; Erin Ivory, "'Sweet Beginnings' Helping Former Inmates Get Bee-keeper Jobs Right Out of Prison," October 11, 2017, Chicago WGN 9, https://wgntv.com/2017/10/11/sweet-beginnings-helping-former -inmates-get-beekeeper-jobs-right-out-of-prison.

7. Author interview with Sarah Nelson Wright, March 23, 2010.

8. "Our History," n.d., Acme Smoked Fish, http://www.acmesmoked fish.com/about-us/our-history; "About Us," n.d., Royal Engraving, Inc., http://www.royalengravinginc.com; Matthew Rivera, "The Art of the Book," *Wall Street Journal*, September 18, 2010, https://www.wsj.com /articles/SB10001424052748703904304575497931119789408.

9. Author interview with Sarah Nelson Wright, March 23, 2010.

10. "What Is 'The Plant'?" February 12, 2016, Plant Chicago, https:// plantchicago.org/2016/02/12/what-is-the-plant.

11. Kari Lydersen, "Chicago Nears End of Era in Stockyards," *Washington Post*, July 18, 2005, http://www.washingtonpost.com/wp-dyn /content/article/2005/07/17/AR2005071700996.html.

12. David King, "Growing Up: How the Plant Is Making Futuristic Farming Today's Reality," February 21, 2013, Newcity, https://newcity .com/2013/02/21/growing-up-how-the-plant-is-making-futuristic -farming-todays-reality; "The Plant," n.d., Bubbly Dynamics, LLC, https://www.bubblydynamics.com/the-plant.

13. Julia Thiel, "The Plant Gets Funding to Move Forward," *Chicago Reader*, September 19, 2011, https://www.chicagoreader.com/Bleader /archives/2011/09/19/the-plant-gets-funding-to-move-forward; author interview with John Edel, February 2, 2019.

14. Matthew Kazin, "U.S. Army Vets Seek to Spice Up Global Saffron Biz," FOXBusiness, July 3, 2016, https://www.foxbusiness.com /features/u-s-army-vets-seek-to-spice-up-global-saffron-biz.

15. Author interview with Kassandra Hinrichsen, February 27, 2018;

Ed Komenda and Tanveer Ali, "Back of the Yards Feel 'Abandoned' by City as Violence Takes Over," DNAinfo, January 6, 2017, https:// www.dnainfo.com/chicago/20170104/back-of-yards/back-of-yards -violence-area-has-been-abandoned.

16. Author interview with Kassandra Hinrichsen, February 27, 2018.

17. Author interview with Kassandra Hinrichsen, February 27, 2018.

18. "About," Bubbly Dynamics, LLC, https://www.bubblydynamics .com/about.

19. Author interview with John Edel, March 24, 2010.

20. Author interview with John Edel, March 24, 2010.

21. Hal Dardick, "City Has $35 Million Backlog in Razing Abandoned Properties," *Chicago Tribune*, October 27, 2014, https://www.chicago tribune.com/news/local/politics/ct-chicago-budget-hearings-1028 -20141027-story.html.

22. Author interview with John Edel, March 24, 2010; author interview with John Edel, February 2, 2019.

23. "Bubbly (aka the Chicago Sustainable Manufacturing Center)," Bubbly Dynamics, LLC, https://www.bubblydynamics.com/chicago-sustainable-manufacturing-center; Registration Form: The Central Manufacturing District: Original East Historic District, December 21, 2015, 15, US Department of the Interior, National Register of Historic Places, https://www.nps.gov/nr/feature/places/pdfs/16000004.pdf.

24. Registration Form, 15; "Bubbly (aka the Chicago Sustainable Manufacturing Center)."

25. "Bubbly (aka the Chicago Sustainable Manufacturing Center)."

26. Upton Sinclair, *The Jungle* (New York: New American Library, 1905), 76; "Bubbly (aka the Chicago Sustainable Manufacturing Center)."

27. Jeremiah Staes, "There's a Reason for All the Hype about Cork-town Detroit," November 28, 2018, Visitdetroit, https://visitdetroit .com/theres-a-reason-for-all-the-hype-about-corktown.

28. Cooley quoted in Danielle Leath, "Detroit Duos: Phil Cooley and Kate Bordine," TBD, February 2017, https://www.tbdmag.com/detroit -duos-phil-cooley-kate-bordine; and "Twenty in Their 20s: Phillip Cooley," *Crain's Detroit Business*, n.d., https://www.crainsdetroit.com /awards/phillip-cooley.

29. Author interview with Phil Cooley, August 9, 2014; Hudson-Webber Foundation, "7.2 SQ MI: A Report on Greater Downtown Detroit," February 2013, 28, 29, https://detroitsevenpointtwo.com/wp-content /uploads/2018/10/2013-Full-Report.pdf.

30. "Our Focus," Ponyride, https://www.ponyride.org/start-your-own; "Ponyride's Mission," Ponyride, https://www.ponyride.org.

31. "Our Focus" (Ponyride); "Ponyride Becomes First Detroit Coworking Space to Offer On-Site Childcare," September 5, 2017, Model D, http://www.modeldmedia.com/inthenews/coworking-childcare -090517.aspx.

32. "Detroit Denim Co. Moving Out of Corktown's Ponyride and into Bigger Digs on Riverfront," January 19, 2016, Model D, http://www .modeldmedia.com/startupnews/detroitdenimcoriverfront011916 .aspx; M. J. Galbraith, "Empowerment Plans Starts Selling New Line of Coats, Preps for Move to Larger West Village Facility," December 19, 2017, Model D, http://www.modeldmedia.com/devnews/Maxwell DetroitDebut.aspx.

33. Galbraith, "Empowerment Plans Starts Selling New Line of Coats."

34. Cooley quoted in Ryan Patrick Hooper, "Ponyride Plans to Sell Corktown Building and Move to Make Art Work Near New Center," *Detroit Free Press*, February 9, 2018, https://www.freep.com/story /entertainment/arts/2018/02/09/ponyride-corktown-move-detroit -make-art-work-recycle-here/321501002; Annalise Frank, "Ford Begins First Phase of $350 Million Makeover of Michigan Central Station," *Crain's Detroit Business*, December 4, 2018, https://www .crainsdetroit.com/real-estate/ford-begins-first-phase-350-million -makeover-michigan-central-station.

35. Kate Opalewski, "Avalon Breads Rises to New Heights," Pride Source, January 31, 2013, https://pridesource.com/article/58218-2; "Countdown to Fun," *Detroit Metro Times*, June 16, 2010, https://www .metrotimes.com/detroit/countdown-to-fun/Content?oid=2197628.

36. Lee DeVito, "Whole Foods Opens in Midtown," *Detroit Metro Times*, June 5, 2013, https://www.metrotimes.com/detroit/whole -foods-opens-in-midtown/Content?oid=2146179.

37. Ryan Felton, "The Activist and Baker: Ann Perrault," *Detroit Metro Times*, June 10, 2015, https://www.metrotimes.com/detroit/the -activist-and-baker-ann-perrault/Content?oid=2349200.

38. Pamela N. Danziger, "How Detroit Denim, Nike, Whole Foods and Shinola Are Rebuilding Retail in Detroit," *Forbes*, March 7, 2018, https://www.forbes.com/sites/pamdanziger/2018/03/07/tour -retails-urban-future-with-whole-foods-shinola-nike-and-detroit -denim-in-detroit/#5.2e23989b5987.

39. Danziger, "How Detroit Denim, Nike, Whole Foods and Shinola Are Rebuilding Retail in Detroit."

40. "A Narrative for the Hope District," n.d., Friends of Detroit and Tri County, http://friendsofdetroit.org/infobox/a-narrative-for-the-hope -district.

41. "A Narrative for the Hope District"; "Club Technology," Friends of Detroit and Tri County, http://friendsofdetroit.org/projects/club -technology; "Hope Gardens," Friends of Detroit and Tri County, http://friendsofdetroit.org/projects/hope-gardens.

42. "About Us," n.d., Friend of Detroit and Tri County, http://friendsof detroit.org/about-us.

43. Author interview with Mike Wemberley, April 28, 2010; John P. Kretzmann and John L. McKnight, *Building Communities from the In- side Out: A Path toward Finding and Mobilizing a Community's Assets* (Chicago: ACTA, 1993).

44. Kretzmann and McKnight, *Building Communities from the Inside Out*, 1, 2; "A Narrative for the Hope District."

45. Grace Lee Boggs, *The Next American Revolution: Sustainable Ac- tivism for the Twenty-First Century* (Berkeley and Los Angeles: Uni- versity of California Press, 2011), 106, 107.

46. Author interview with Mike Wemberley, April 28, 2010.

47. Author interview with Mike Wemberley, April 28, 2010.

48. Susan Tompor, "With Boost from Oprah Winfrey, Detroit Friends Potato Chips Hit It Big," *Detroit Free Press*, November 7, 2016, https:// www.freep.com/story/money/personal-finance/susan-tompor/2016 /11/07/oprah-gives-blessing-detroit-friends-potato-chips/9342 8520.

49. Author interview with Mike Wimberley, April 28, 2010; Kurt Nagl, "$1 Million Grant to Help Detroit Creative Corridor Move ahead with City Design Initiative," *Crain's Detroit Business*, February 16, 2017, https://www.crainsdetroit.com/article/20170216/NEWS/170219889 /1-million-grant-to-help-detroit-creative-corridor-center-move -ahead; Aaron Mondry, "Green Garage Turns Five: How a Business Incubator Created a Lasting Model for Sustainability," June 13, 2016, Model D, http://www.modeldmedia.com/features/greengarage-five -sustainability-061116.aspx.

50. Bill Wylie-Kellermann, "Reading Rivera: Resurrection and Re- membering in Post-Industrial Detroit," *The Record*, December 2009, https://skipschiel.files.wordpress.com/2014/04/reading-rivera-bill -wylie-kellerman.pdf; "Detroit's Unemployment Rate Is Nearly 50%, according to the *Detroit News*," *Huffington Post*, May 25, 2011, https:// www.huffpost.com/entry/detroits-unemployment-rat_n_394559.

52. Wylie-Kellermann, "Reading Rivera."

52. Jonathan Oosting, "Detroit Has More Vacant Land Than Any City in Nation Except Post-Katrina New Orleans," September 29, 2009, M Live, https://www.mlive.com/news/detroit/2009/09/detroit_has _more_vacant_land_t.html; Alexandra Alter, "Artists vs. Blight," *Wall Street Journal*, April 17, 2009, https://www.wsj.com/articles/SB12399231 8352327147; Allison Plyer, "Population Loss and Vacant Housing in New Orleans Neighborhoods," February 5, 2011, The Data Center, https://www.datacenterresearch.org/reports_analysis/population -loss-and-vacant-housing; author interview with Bill Wylie-Keller- mann, October 27, 2010; Nina Ignaczak, "No Stranger to Urban Agriculture, Detroit Makes It Official with New Zoning Ordinance," Seedstock, April 9, 2013, http://seedstock.com/2013/04/09/no-stranger-to-urban-agriculture-detroit-makes-it-official-with-new-zoning-ordinance.

53. Author interview with Bill Wyle-Kellermann, October 27, 2010.

54. Grace Lee Boggs, "My Philosophic Journey," *The Boggs Blog*, April 5, 2014, https://conversationsthatyouwillneverfinish.wordpress.com/2014/04/05/my-philosophic-journey-by-grace-lee-boggs; Mark Engler, "From the U.S. Social Forum," *Dissent*, June 23, 2010, https://www.dissentmagazine.org/blog/from-the-u-s-social-forum.

55. Gus Burns, "23 Arrested for Blocking Detroit Streetcar, Building Entrance," M Live, June 19, 2018, https://www.mlive.com/news/detroit/2018/06/23_protesters_arrested_for_blo.html; "Gilbert 7 Press Conf Notes," document in authors' possession; correspondence with Bill Wylie-Kellermann, March 1, 2019. In February 2019, Wylie-Kellermann was acquitted of all charges stemming from his June 2018 arrest.

56. Andy Tarnoff, "The Empty Shops of the Not-So-Grand Avenue," November 30, 2010, OnMilwaukee, https://onmilwaukee.com/market/articles/grandavedecline.html; Dave Reid, "The Harley-Davidson Museum Celebrates an American Icon," July 12, 2008, Urban Milwaukee, https://urbanmilwaukee.com/2008/07/12/the-harley-davidson-museum-celebrates-an-american-icon; David Goldman, "Bad Quarter? Time to Swing the Job Ax," April 19, 2008, CNN, https://money.cnn.com/2008/04/18/news/companies/job_cuts/index.htm?postversion=2008041812.

57. Author interview with Michael Frede, February 9, 2010.

58. Tom Daykin, "Community Warehouse to Open North Ave. Store," *Milwaukee Journal-Sentinel*, January 3, 2017, https://www.jsonline.com/story/money/real-estate/commercial/2017/01/03/community-warehouse-open-north-ave-store/96104528.

59. Author interview with George Bogdanovich, May 6, 2010.

60. Community Warehouse, https://www.thecommunitywarehouse.org.

61. Author interview with Jacob Maclin, May 6, 2010; author interview with Rodrigo Cisneros, May 6, 2010.

62. Author interview with Nick Rinnger, March 16, 2018.

63. Author interview with Nick Rinnger, March 16, 2018; author interview with Adam Procell, January 21, 2020.

64. Author interview with Amy Kaherl, March 9, 2018; David Sands, "Every Dollar Counts: Detroit SOUP and the Impact of Micro-Grants," April 17, 2017, Model D, http://www.modeldmedia.com/features/detroit-soup-impact-041717.aspx.

65. Author interview with Amy Kaherl, March 9, 2018.

66. Author interview with Amy Kaherl, March 9, 2018.

67. Author interview with Amy Kaherl, March 9, 2018.

Chapter 6

1. See, e.g., the Indiegogo campaign for "Leaving Our Mark in Lake Park: A Mural Project 2": https://www.indiegogo.com/projects/leaving-our-mark-in-lake-park-a-mural-project-2-art-community#6.

2. Markusen and Gadwa, "Creative Placemaking," 3.

3. For details on work for pay, see Work for Pay, https://workforpay.wordpress.com.

4. Author interview with Lydia Bell, March 18, 2010.

5. Author interview with Lydia Bell, March 18, 2010.

6. For information on The Work Office (TWO), see http://www.theworkoffice.com/html/about.html; author interview with Lydia Bell, August 27, 2018.

7. Much of FEAST Brooklyn's work can be accessed through http://feastinbklyn.org.

8. Author interview with Jeff Hnilicka, March 19, 2018.

9. Author interview with Jeff Hnilicka, March 19, 2018.

10. Author interview with Jeff Hnilicka, March 19, 2018.

11. Author interview with Melanie Jelacic, March 16, 2010; author interview with Melanie Jelacic, August 16, 2018.

12. For Funderburgh's FEAST project, see "Winner! Dan Funderburgh," February 21, 2009, FEAST Brooklyn, http://feastinbklyn.org/?p=31; author interview with Dan Funderburgh, March 25, 2010.

13. "Winner! Trinity Project," April 24, 2010, FEAST Brooklyn, http://feastinbklyn.org/?p=344.

14. Correspondence with Monica Salazar Olmsted, March 8, 2019.

15. Correspondence with Monica Salazar Olmsted, March 8, 2019.

16. Lynn Maliszewski, "Peter Clough's One and Three Quarters of an Inch," *Whitehot Magazine*, October 2010, https://whitehotmagazine.com/articles/one-three-quarters-an-inch/2144; correspondence with Peter Clough, February 21, 2019.

17. For Dan Funderburgh's work, see "About," n.d., Dan Funderburgh, https://cargocollective.com/DFgamma/ABOUT-CV; author interview with Dan Funderburgh, March 19, 2018.

18. Alex Greenburger, "Trump Proposes Cuts to NEA, NEH Funding in 2019 Budget," *ArtNews*, February 12, 2018, https://www.artnews.com/art-news/news/trump-administration-proposes-cuts-nea-neh-funding-2019-budget-9800.

19. "Milton Mizenberg and His Oakland Museum of Contemporary Art," *Detour Art*, May 7, 2009, http://www.detourart.com/milton-mizenberg-and-his-oakland-museum-of-contemporary-art; "Hamtramck

Disneyland," n.d., Hatch Art, http://www.hatchart.org/hamtramck -disneyland.

20. "About Philadelphia's Magic Gardens," n.d., Philadelphia's Magic Gardens, https://www.phillymagicgardens.org/about-us.

21. "History," n.d., Heidelberg Project, https://www.heidelberg.org /history.

22. "Timeline," n.d., Heidelberg Project, https://www.heidelberg.org /timeline; "The Heidelberg Arts Leadership Academy (HALA) Is to Launch in January, 2018," Heidelberg Project, https://www.heidelberg .org/recent-news/2017/11/28/the-heidelberg-arts-leadership -academy-to-launch-in-2018-with-your-support; Lee DeVito, "Heidelberg Project Announces New Headquarters Plans," *Detroit Metro Times*, September 21, 2018, https://www.metrotimes.com/the-scene /archives/2018/09/21/heidelberg-project-announces-new-head quarters-plans.

23. "When Artists Break Ground: Lessons from a Cleveland Neighborhood Project" (Cleveland: Community Partnership for Arts and Culture and Northeast Shores Development Corporation, 2014), https:// kresge.org/sites/default/files/report-Northeast-Shore-Cleveland -2014.pdf.

24. John Asante, "Peek Inside Copy Cat Building: Where Baltimore Artists Work—and Live," March 22, 2012, NPR, https://www.npr.org /sections/pictureshow/2012/03/22/149061691/peek-inside -the-copy-cat-building-where-baltimore-artists-work-and-live; Victor Bensus, Irene Calimlin, Anna Cash, Scott Chilberg, Reshad Hai, Eli Kaplan, and Aline Tanielian, "Punitive to Rehabilitative: Strategies for Live-Work Preservation in Oakland," Community Development Studio Final Project (Berkeley: University of California, Berkeley, Department of City and Regional Planning, Spring 2018), 27, http://www .urbandisplacement.org/sites/default/files/images/livework_ucb _studio_report_final.pdf.

25. Rick Paulas, "The Quasi-Conspiracy That Caused Oakland's Deadly Ghost Ship Fire," *Huffington Post*, November 5, 2018, https:// www.huffingtonpost.com/entry/oakland-ghost-ship-fire_us_5bdcbc 61e4b01ffb1d02381d.

26. Kinsee Morlan, "Another Arts Venue Bites the Dust," August 23, 2017, Voice of San Diego, https://www.voiceofsandiego.org/topics /arts/another-arts-venue-bites-dust.

27. "History," n.d., AS220, https://as220.org/about/history.

28. "History" (AS220); "Sponsors," n.d., AS220, https://as220.org/about /sponsors; "Calendar," AS220, https://as220.org/calendar.

29. "Unpacking Authentic Placemaking," October 1, 2015, AS220, https://as220.org/unpacking-authentic-placemaking-october-6th.

30. Jay Gabler, "Bon Iver Announce '22, a Million' Listening Parties around the World," *The Current*, September 29, 2016, https://blog .thecurrent.org/2016/09/bon-iver-announce-22-a-million-listening -parties-around-the-world.

31. Eric Avila, *The Folklore of the Freeway: Race and Revolt in the Modernist City* (Minneapolis: University of Minnesota Press, 2014), 160–71.

32. Avila, *The Folklore of the Freeway*, 167–70.

33. Quevedo quoted in Kevin Delgado, "A Turning Point," *Journal of San Diego History*, vol. 44, no. 1 (Winter 1998), http://www.sandiego history.org/journal/1998/january/chicano-3.

34. "Barrio Logan Historical Resources Survey" (San Diego: City of San Diego, City Planning and Community Investment, Community Planning and Urban Forms Division, February 1, 2011), https://www .sandiego.gov/sites/default/files/legacy/planning/programs/his torical/pdf/2013/201304blhistoricsurvey.pdf; "Barrio Logan Community Plan and Local Coastal Program" (San Diego: City of San Diego, 2013), https://www.sandiego.gov/sites/default/files/legacy /planning/community/cpu/barriologan/pdf/barrio_logan_cpu_full _090313.pdf; Gary Warth, "Chicano Park Named National Historic Landmark," *San Diego Union Tribune*, January 11, 2017, https://www .sandiegouniontribune.com/news/politics/sd-me-chicano-historic -20170111-story.html; Chase Scheinbaum, "Celebrate Chicano Park Day in San Diego This Weekend," *San Diego Magazine*, April 18, 2018, https://www.sandiegomagazine.com/Blogs/Cityfiles/Spring-2018 /Celebrate-Chicano-Park-Day-in-San-Diego-This-Weekend.

35. For a general history of the Philadelphia Mural Arts Program, see Jane Golden, Robin Rice, and Monica Yant Kinney, *Philadelphia Murals and the Stories They Tell* (Philadelphia: Temple University Press, 2002).

36. "Mural Arts Philadelphia Is the Nation's Largest Public Art Program, Dedicated to the Belief That Art Ignites Change," n.d., Press kit, Mural Arts Philadelphia, https://www.muralarts.org/wp-content/up loads/2018/01/MA_Press-Kit_FINAL_January2018.pdf; "The Guild," n.d., Mural Arts Philadelphia, https://www.muralarts.org/program /restorative-justice/the-guild.

37. Silvana Pop, "When Silenced Voices Become Wall-Scale Public Discourse: A Social Constructionist Analysis of Two Philadelphia Murals," *CONCEPT*, vol. 35 (2012), https://pdfs.semanticscholar.org /9825/87ca1fb2f8151912ea36703c7f91fe30cd5f.pdf.

38. Nicole Jones, "Rejuvenating the Community, One Mural at a Time," August 12, 2011, OaklandNorth, https://oaklandnorth.net/2011 /08/12/rejuvenating-the-community-one-mural-at-a-time.

39. Jones, "Rejuvenating the Community, One Mural at a Time."

40. Melero quoted in the video "Bringing Funk to Funktown," January 4, 2010, https://www.youtube.com/watch?v=cxkdgtRG7cE; Ben

Valentine, "Group behind Oakland's Newest Mural Uses Walls to Depict Communities," KQED, August 7, 2014, https://www.kqed.org/arts/10139706/oaklands-newest-mural.

41. For more on the Living Walls conference, see the event's webpage: https://www.livingwallsatl.com/living-walls-conference-1.

42. Hugh Leeman, "MLK vs. Old English: A Letter from Living Walls Artist Hugh Leeman," August 12, 2010, Creativeloafing, https://creativeloafing.com/content-212374-MLK-vs.-Old-English:-A-letter-from-Living-Walls-artist-Hugh-Leeman.

43. John Schmid, "Still Separate and Unequal: A Dream Derailed," *Milwaukee Journal-Sentinel*, December 5, 2004; Karalee Surface, "A. O. Smith Corporation," n.d., *Encyclopedia of Milwaukee*, https://emke.uwm.edu/entry/a-o-smith-corporation; "Century City Business Park," n.d., City of Milwaukee, Department of City Development, https://city.milwaukee.gov/CenturyCity.

44. David Lapeska, "The Midwest's Rebuild Era, in Pictures," May 15, 2012, CityLab, https://www.citylab.com/design/2012/05/observing-midwest-rebuild/2004.

45. Chris King, "Painter of the People: Chris Green Is Beautifying North City, One Abandoned Building at a Time," *St. Louis American*, September 15, 2016, http://www.stlamerican.com/news/local_news/painter-of-the-people-chris-green-is-beautifying-north-city/article_2a44a37c-7adc-11e6-9824-1b8d7f088e0a.html.

46. Author interview with Dennis Maher, April 11, 2011.

47. Sara Maurer, "A Closer Look at the Fargo House: A Center for Urban Imagination," February 3, 2013, Buffalo Rising, https://www.buffalorising.com/2013/02/a-closer-look-at-the-fargo-house-a-center-for-urban-imagination.

48. "AK Innovation Lab's SACRA Project Recognized with NEA Our Town Grant," Press release, June 15, 2017, Albright-Knox, https://www.albrightknox.org/news/ak-innovation-lab%E2%80%99s-sacra-project-recognized-nea-our-town-grant.

49. "The Society for the Advancement of Construction-Related Arts," Albright Knox, https://www.albrightknox.org/community/ak-innovation-lab/society-advancement-construction-related-arts-sacra.

50. "Our Mission," n.d., Green Project, https://www.thegreenproject.org/mission-story; "Salvations Gala + Auction," n.d., Green Project, https://www.thegreenproject.org/salvations.

51. "Mission & History," n.d., Scroungers' Center for Reusable Art Parts (SCRAP), https://sites.google.com/scrap-sf.org/home/about-scrap/mission-history.

52. Author interview with Shauta Marsh, June 14, 2018.

53. Author interview with Jim Walker, June 10, 2018.

54. Author interview with Jim Walker, June 10, 2018.

55. Author interview with Shauta Marsh, June 14, 2018.

56. Author interview with Jim Walker, June 10, 2018.

57. Samantha Michaels, "A Conversation with Candy Chang, Public Installation Artist and Designer," *The Atlantic*, August 15, 2011, https://www.theatlantic.com/national/archive/2011/08/a-conversation-with-candy-chang-public-installation-artist-and-designer/243630.

58. Author interview with Candy Chang, April 12, 2011; "I Wish This Was," n.d., Candy Chang, http://candychang.com/work/i-wish-this-was.

59. Author interview with Candy Chang, April 12, 2011. Numerous other sites were also visited.

60. Julia Thiel, "The Opportunity Shop Opens in Hyde Park," *Chicago Reader*, November 27, 2009, https://www.chicagoreader.com/Bleader/archives/2009/11/27/the-opportunity-shop-opens-in-hyde-park.

61. Britany Robinson, "Laura Shaeffer Talks Op Shop," March 19, 2010, *Gapers Block*, http://gapersblock.com/ac/2010/03/19/laura-shaeffer-talks-op-shop.

62. Author interview with Laura Shaeffer, March 5, 2019.

63. Author interview with Laura Shaeffer, June 10, 2018.

64. Rachel Cromidas, "In Grand Crossing, a House Becomes a Home for Art," *New York Times*, April 7, 2011, https://www.nytimes.com/2011/04/08/us/08cncculture.html?mtrref=www.google.com&gwh=AA16211E58B3941630D4D978BBEE38C1&gwt=pay&assetType=REGIWALL.

65. Public conversation with Theaster Gates at the University of Chicago, May 1, 2012.

66. Public conversation with Theaster Gates at the University of Chicago, May 1, 2012.

67. Cromidas, "In Grand Crossing, a House Becomes a Home for Art."

68. Maudlyne Ihejirika, "Dedication Held at New South Side Home for Cleveland Gazebo Where Tamir Rice Was Killed," *Chicago Sun-Times*, June 23, 2019, https://chicago.suntimes.com/2019/6/23/18714853/dedication-south-side-arts-bank-cleveland-gazebo-tamir-rice-killed.

69. Alan G. Brake, "CHA's New Way," *Architect's Newspaper*, November 9, 2011, https://archpaper.com/2011/11/chas-new-way.

70. Public conversation with Theaster Gates at the University of Chicago, May 1, 2012.

71. Lee DeVito, "Finally, Detroit's RoboCop Statue Has a Home," *Detroit Metro Times*, May 2, 2018, https://www.metrotimes.com/the-scene/archives/2018/05/02/finally-detroits-robocop-statue-has-a-home; "Detroit Needs a Statue of Robocop!," Project Update,

Kickstarter.com, December 31, 2019, https://www.kickstarter.com /projects/imaginationstation/detroit-needs-a-statue-of-robocop /posts; Julie Hinds, "Detroit's RoboCop Statue, a Magnificent Obsession for 9 Years Now, Is Nearly Done," *Detroit Free Press*, January 14, 2020, https://www.freep.com/story/entertainment/movies/julie-hinds /2020/01/14/robocop-statue-detroit/4456086002/.

72. Author interview with Jerry Paffendorf, October 23, 2011.

73. Paffendorf quoted in Melena Ryzik, "Wringing Art Out of the Rubble in Detroit," *New York Times*, August 3, 2010, https://www.nytimes .com/2010/08/04/arts/design/04maker.html.

74. Paffendorf quoted in Ryzik, "Wringing Art Out of the Rubble in Detroit"; Travis R. Wright, "Raze One, Raise One," *Detroit Metro Times*, July 28, 2010, https://www.metrotimes.com/detroit/raze-one-raise -one/Content?oid=2197860.

75. Wright, "Raze One, Raise One"; Lisa Smith, "Catie Newhall's Salvage Landscape Reclaims an Arsoned Building in Detroit," November 2, 2010, Core 77, https://www.core77.com/posts/17772/cat ie-newells-salvaged-landscape-reclaims-an-arsoned-building-in-de troit-17772.

76. Information on the sale of Righty taken from the Imagination Station Facebook page: https://www.facebook.com/facethestation.

77. Sarah Rose Sharp, "Building Monuments amid Detroit's Modern Day Ruins," May 11, 2015, Hyperallergic, https://hyperallergic.com /205716/building-monuments-amid-detroits-modern-day-ruins.

78. For a history of Project Row Houses, see Ryan N. Dennis, ed., *Collective Creative Actions: Project Row Houses at 25* (Durham, NC: Duke University Press, 2018).

79. "Architecture," n.d., Project Row Houses, https://projectrowhouses .org/architecture; "Project Row Houses Now Accepting Applications for Young Mothers Residential Program," August 22, 2018, CW39 Houston, https://cw39.com/2018/08/22/project-row-houses-now -accepting-applications-for-young-mothers-residential-program; Kriston Capps, "How a Houston Housing Project Earned a MacArthur Grant," September 19, 2014, CityLab, https://www.citylab.com/design /2014/09/how-a-houston-housing-project-earned-a-macarthur -grant/380401.

80. Small group conversation with Rick Lowe, October 24, 2015.

81. Small group conversation with Rick Lowe, October 24, 2015.

82. Small group conversation with Rick Lowe, October 24, 2015.

83. Mike Hixenbaugh, "Third Ward Leaders Want to Preserve Community's African American Legacy," *Houston Chronicle*, January 13, 2019, https://www.houstonchronicle.com/news/houston-texas/houston /article/Third-Ward-leaders-work-to-preserve-community-s-135 30293.php.

Conclusion

1. Sue Halpern, "Mayor of Rust," *New York Times*, February 11, 2011, https://www.nytimes.com/2011/02/13/magazine/13Fetterman-t .html.

2. Halpern, "Mayor of Rust"; Tom MacDonald, "Rendell Backs Fetterman Campaign for Pa. Lieutenant Governor," December 4, 2017, WHYY, https://whyy.org/articles/rendell-backs-fetterman-campaign -pa-lieutenant-governor.

3. Halpern, "Mayor of Rust"; "About UnSmoke," n.d., UnSmoke Systems Artspace, http://unsmokeartspace.com/about.

4. "$153,000 in Grants to Support Local Artists," n.d., Pittsburgh Foundation, https://pittsburghfoundation.org/node/38722; Into the Furnace, https://intothefurnace.wordpress.com.

5. "Braddock Farms," n.d., Grow Pittsburgh, https://www.growpitts burgh.org/about-us/locations/braddock-farms.

6. "Braddock Farms"; "Job Training Program," Superior Motors, https://www.superiormotors15104.com.

7. Leah Lizarondo, "Gisele Barreto Fetterman: The Free Store Celebrates Two Years," September 29, 2014, Next Pittsburgh, https://www .nextpittsburgh.com/next-wave/gisele-barreto-fetterman-freestore -celebrates-two-years.

8. Kim Lyons, "Braddock Unveils Renovated Free Press Building, Studebaker Metals Opens Within," October 30, 2015, Next Pittsburgh, https:// www.nextpittsburgh.com/city-design/braddock-unveils-renovated -revitalized-free-press-building.

9. Max Graham, "Coming Full Circle: Town Square Opens at Former Braddock Hospital," *Pittsburgh Post-Gazette*, August 10, 2018, https://www.post-gazette.com/local/east/2018/08/10/Coming-full -circle-Town-square-opens-at-former-Braddock-hospital/stories /201808060003.

10. John Powers, "A Look at the Creative Time Summit: Gentrification, Gentrification, and Gentrification," November 5, 2013, ArtFCity, http://artfcity.com/2013/11/05/a-look-at-the-creative-time-summit- gentrification-gentrification-and-gentrification.

11. "Braddock Marks Fifth Anniversary without Homicide as Crime Rates Fall," May 24, 2013, WPXI 11 News, https://www.wpxi.com /news/local/braddock-marks-fifth-anniversary-without-homicide -/289425912.

12. Jim Straub and Bret Liebendorfer, "Braddock, Pennsylvania Out of the Furnace and into the Fire," *Monthly Review*, December 1, 2008, https://monthlyreview.org/2008/12/01/braddock-pennsylvania-out -of-the-furnace-and-into-the-fire.

13. Christine H. O'Toole, "Braddock Rising," *Pittsburgh Magazine*,

March 25, 2015, https://www.pittsburghmagazine.com/braddock
-rising; Morgan quoted in Halpern, "Mayor of Rust."

14. "Ready to Work" (Levi's ad), https://www.youtube.com/watch?v=
oURB0C0vTel.

15. Kelly Klaasmeyer, "Love among the Ruins: A Town in Decline in
LaToya Ruby Frazier's Witness," July 24, 2013, Houston Press, https://
www.houstonpress.com/arts/love-among-the-ruins-a-town-in-de
cline-in-latoya-ruby-fraziers-witness-6598802.

16. Author interview with Isaac Bunn, September 20, 2018.

17. Author interview with Isaac Bunn, September 20, 2018.

18. Author interview with Isaac Bunn, September 20, 2018.

19. Author interview with Isaac Bunn, September 20, 2018.

20. Kieran Mclean, "Community Speaks Out against Mercy Hospital Expansion," *The Pitt News*, July 18, 2018, https://pittnews.com
/article/133579/featured/community-speaks-out-against-mercy
-hospital-expansion.

21. Sarah Hauer, "Will Allen retiring as Growing Power alters course,"
Milwaukee Journal-Sentinel, November 20, 2017, https://www.jsonline
.com/story/money/business/2017/11/20/growing-power-founder
-allen-retire-nonprofits-debts-mount/881139001.

22. "Growing Power Awarded $5 Million Grant to Grow Commu-
nity Food Projects across the U.S.," Press release, September 8, 2012,
W. K. Kellogg Foundation, https://www.wkkf.org/news-and-media
/article/2012/09/growing-power-awarded-$5-million-grant.

23. Stephen Satterfield, "Behind the Rise and Fall of Growing Power,"
Civil Eats, March 13, 2018, https://civileats.com/2018/03/13/behind
-the-rise-and-fall-of-growing-power.

24. Satterfield, "Behind the Rise and Fall of Growing Power."

25. Growing Power promotional materials, in the authors' possession;
White, *Freedom Farmers*, 141; Bolden-Newsome and Cook quoted in
Leah Penniman, "After a Century in Decline, Black Farmers Are Back
and on the Rise," *Yes! Magazine*, May 5, 2016, https://www.yesmagazine
.org/people-power/after-a-century-in-decline-black-farmers-are
-back-and-on-the-rise-20160505.

26. Author interview with Dasia Harmon, August 21, 2018.

27. Author interview with Dasia Harmon, August 21, 2018; Tom Las-
kawy, "FoodCorps Will Teach Kids, Link Farms and Schools," August 16,
2011, Grist, https://grist.org/sustainable-food/2011-08-15-foodcorps
-launches-will-teach-kids-link-farms-schools.

28. Author interview with Dasia Harmon, August 21, 2018.

29. Mary Sussmann, "Will Allen Returns to His Roots," June 12, 2018,
Shepherd Express, https://shepherdexpress.com/news/features/will
-allen-returns-to-his-roots/#/questions.

Index

ACME Collective, 78

Acme Smoked Fish, 159

activism, 186; agriculture and, 114–18, 121, 141, 151; art and, 243; Black Panther Party and, 121; community and, 271–72, 278; contemporary placemaking and, 52; Growing Power and, 118, 130–51, 155, 273–78; historical perspective and, 4, 14–15, 18–19, 21, 24; innovation of, 4; Jacobs and, 2 (*see also* Jacobs, Jane); NAFTA and, 78; 1990s and, 72, 78; Occupy Wall Street and, 107, 156; production and, 155, 159, 181, 185; protests and, 14, 21, 78, 117; twenty-first century and, 107; Whittier Elementary School and, 15, 18–19

Adams, Larry, 136–37

Adams, Sharon, 136–37, 140

Adams-Hapner, Kerry, 102

affluence, 32, 50, 57, 179, 199

African American Chamber of Commerce, 140

African Americans: agriculture and, 115, 117–18, 121, 124–26, 134–35, 140; art and, 222, 227, 243, 246, 250, 257; Black Panther Party and, 121; Chicago and, 52, 76, 78, 227, 243, 246; community and, 271–74, 278; deindustrialization and, 52, 125, 179; Detroit and, 6, 179; Milwaukee and, 115, 117–18, 121, 124–26, 135, 140, 189, 222; New York and, 52, 72, 243; 1990s and, 61, 72, 76, 78; police tactics and, 72; production and, 179, 189; Project Row Houses and, 199, 211, 257–61, 265, 277; public space and, 61

After School Matters, 76, 78–79

Agee, James, 1

agriculture: activism and, 114–18, 121, 141, 151; African Americans and, 115, 117–18, 121, 124–26, 134–35, 140; Will Allen and, 118–19, 130–31, 134–38, 141–43, 145, 155, 273–74, 276–78; Chicago and, 121, 127, 134–35, 142–43; City Garden Plot Program and, 127; civil rights and, 118, 134–35, 143, 151; Cleveland and, 121, 123, 127, 145, 150; Commercial Urban Agriculture Training Program (CUA) and, 118, 134, 143, 145; composting and, 130, 134, 138, 143, 274–75; creative class and, 113–14, 130, 134, 151; deindustrialization and, 155; Detroit and, 121, 127, 134, 143, 150; education and, 115, 126–29, 134, 140, 142; entrepreneurialism and, 115, 128, 131, 140, 151, 158; extension agents and, 125–27; Florida and, 113–14, 130, 134, 136, 151; 4–H clubs and, 125–26, 129; funding and, 117–18, 127–31, 142, 145, 150; Good Food Revolution and, 118; government and, 113–18, 125–26, 131, 135–36, 151; grants and, 118, 127, 135–36, 142–43, 150; Growing Power and, 118, 130–51, 155, 273–78; Grow Pittsburgh and, 270; growth and, 114, 121, 124, 129–30, 138, 142; historical perspective on, 119–25; housing and, 117, 130, 142–45; industry and, 113–15, 121–25, 134, 138, 141–42, 150–51; investment and, 113, 118, 142, 145; Jacobs and, 114–18, 131, 134, 151; manufacturing and, 121, 130, 138, 151; markets and, 92–93, 102, 104, 123–31, 135, 138, 140, 145, 166, 174, 176, 178, 237, 276; Milwaukee and, 115–51, 155, 273–78; New York and, 114, 127, 130; Partners for Places and, 118; police and, 117, 125, 130, 141; postindustrial economy and, 114–15, 121, 134, 138, 151, 155, 158, 169; poverty and, 117, 125, 150; production and, 155–61, 166, 169; public space and, 123, 155; revitalization and, 2; rooftop gardens and, 140, 155, 198; Shoots 'n Roots and, 125, 127–31; suburbs and, 113, 123–24, 138; unemployment and, 125–26, 128; urban renewal and, 114, 125, 151; Urban Work Group and, 126; USDA and, 105, 125, 127, 135, 166; Verdin and, 115, 118; victory gardens and, 125; Walnut Way and, 136–41, 143, 277; Venice Williams and, 118; zoning and, 136

Airbnb, 156

Albright-Knox Innovation Lab, 232

Alexander, Christopher, 5, 39–40, 47, 67

Alice's Garden, 118

Alinsky, Saul, 52

Allen, Erika, 274

Allen, Will: on aesthetics, 130; civil rights and, 118, 134–35, 143; community and, 277–78; FoodCorps and, 276; Mansfield Frazier and, 145, 150; Good Food Revolution and, 118; Great Migration and, 135; Growing Power and, 118–19, 130–31, 134–38, 141–43, 145, 155, 273–74, 276–78; Partners for Places and, 118; power of place and, 127; Pratt and, 142; Verdin and, 119; Walnut Way and, 136–38, 277; Will's Roadside Farm and Markets and, 276; Yakini and, 143

42, 48, 70, 75, 96; Detroit and, 2, 12; *The Economy of Cities*, 36; Florida and, 2, 37, 40–41, 87–88, 91, 106, 114, 134, 198; Greenwich Village and, 65; influence of, 2, 5, 21; manufacturing and, 36–37; 1990s and, 60, 62, 65, 68–70, 75; Oldenburg and, 49; planning theory and, 36; postwar era and, 101; public space and, 34–35, 40–42, 48–52, 70, 91; rebirth of planning and, 62; *Reconsidering Jane Jacobs* and, 114; Sennett and, 5, 40–41, 49, 51; social capital and, 38, 75; twenty-first century and, 87–88, 91, 96, 101, 106; urban revival and, 34–38
Janowitz, Morris, 94
Jelacic, Melanie, 200–201
Jerinic, Katarina, 199
John D. and Catherine T. MacArthur Foundation, 118
Johnson-Sabir, JoAnne, 140
Juice Kitchen, 138, 140
Jung, Kimberly, 165
Jungle, The (Sinclair), 169

Kaherl, Amy, 189–93
Karina Frost and the Banduvloons, 213
Kay, Ken, 68
Kelling, George L., 42–43, 50, 52, 70
Kennedy-King College, 156–57
Kent, Fred, 1–2, 46
Kickstarter, 156, 200, 249
Kinder Foundation, 4
King, Martin Luther, Jr., 52, 117–18, 185, 220–21
Klose, Alex, 125
Knight Foundation, 104
Knoxville, Johnny, 272
Knuckles, Frankie, 246
Kombuchade, 161
Kool and the Gang, 61
Kresge Foundation, 4, 103–5, 209, 211
Kretzman, John P., 180
Kunstler, James Howard, 67

La Casita, 15, 18–19, 21, 24, 278
Landesman, Rocco, 99–100

Landon Bone Baker Architects, 246
land value tax, 113
Langdon, Philip, 66–67
Lankford, Charles, 209–10
Lao Family Community Development, 220
Lasky, Mary Ann, 70
Latent Design, 19
Latinx neighborhoods, 12, 21, 29, 72, 166, 188, 212–13, 218
Leadership in Energy and Environmental Design (LEED), 15
Lee, Summer, 273, 278
Leeman, Hugh, 221
Lefty, 252
Leonard, A. G., 168–69
Lerman, Liz, 76
Lerner, Max, 30
Levi's, 271–72
Levy, Marilyn W., 46
liberalism, 4, 18, 59, 73, 90, 100, 115, 174, 261, 272, 278
libraries: Angonese and, 14–15; art and, 242; BiblioTreka, 18; Carnegie, 266, 270–71; community and, 13–15, 18, 242, 266, 270–71, 278; do-it-yourself (DIY), 14; fighting for, 13; La Casita, 15, 18–19, 24, 278; Read/Write Library, 15, 18–19
Lindsay Heights, 136–38, 140
Lip Bar, 174
Little Village, 13–14
livestock, 130, 161
"Living Walls, the City Speaks" conference, 221–22
LJ's Lounge, 171
Lloyd, Richard, 93–94, 96
loans, 142, 156, 167
Local Government Commission, 64
Lonely Crowd, The (Riesman), 30
Lore, Matthew, 83
Los Angeles, 60–61, 68, 72–73, 89, 102–3, 127, 258
Louima, Abner, 72
Loveland Technologies, 252
Lowbo, Viejo, 213, 218
Lowe, Rick, 211, 257–61
Lowe Brothers Paint, 167

Lower Ninth Ward, 237
Lukaszekski, Dennis, 131
Lynch, Kevin, 38–39, 42, 70

MacArthur Foundation, 257, 261, 274
Maclin, Jacob, 189
Magic Gardens, 204–5
Maher, Dennis, 227, 232–34, 239, 260
Maier, Henry, 124, 127–28, 130
Make Art Work, 176
Make Chicago, 168
Make It Right Foundation, 237
Mantyh, Mary, 129
manufacturing: agriculture and, 121, 130, 138, 151; art and, 223, 234; Brooklyn and, 159–60; Chicago and, 160–61, 166, 168–69; Cleveland and, 104, 277; community and, 271, 273, 277; contemporary placemaking and, 36–37, 43, 52; Detroit and, 90, 104, 169, 171, 174, 178, 192, 277; Jacobs on, 36–37; Milwaukee and, 109, 121, 138, 151, 187, 223; New York and, 43; 1990s and, 58, 61, 67; postwar era and, 29; production and, 158–60, 166–71, 174, 178, 187, 192; twenty-first century and, 87, 90, 94, 97, 104, 109
Markusen, Ann, 3–4, 12, 21, 100–106, 134, 198
Marsh, Shauta, 235
Martin Luther King Jr. National Historical Park, 221
Martin Luther King Jr. Peace Place, 117–18
Martin Luther King Way, 220
Maryland Institute College of Art, 210
Mason, Biddy, 60
Massachusetts Institute of Technology (MIT), 3, 260
Massachusetts Museum of Contemporary Art (MASS Moca), 62
Maxwell Detroit, 174, 278
May, Derrick, 1
May, Elaine Tyler, 30
McGee, Stephen, 252
McKnight, John L., 180
Medukic, Ivana, 209
Mellon Foundation, 104

Index

Plater-Zyberk, Elizabeth, 64–67

police: agriculture and, 117, 125, 130, 141; arrests and, 18, 39, 72, 179, 186; art and, 220, 246; broken-window theory and, 42, 70; brutality and, 72, 107, 117, 220, 246, 260; community and, 42, 70, 117; contemporary placemaking and, 42–43, 47, 50; frisking and, 72; New York and, 70, 73; 1990s and, 58, 70, 72; production and, 179; public space and, 42, 48, 50, 70, 72; race and, 72, 179; twenty-first century and, 107

Polyzoides, Stefanos, 64

Ponyride Market Summer Series, 174, 176, 192, 277

Poor People's Campaign, 185–86

postindustrial economy, 5; agriculture and, 114–15, 121, 134, 138, 151, 155, 158; art and, 198; Chicago and, 15, 57, 94, 168, 198; contemporary placemaking and, 53; Detroit and, 6, 57; Florida and, 158; Milwaukee and, 115, 121, 138, 155, 187; National Endowment for the Arts (NEA) and, 158; New York and, 57, 114, 155, 159; 1990s and, 57–58, 61; production and, 155, 158–78, 187; twenty-first century and, 86, 94, 97, 104

postrecession economy, 6, 86, 159, 233

postwar era, 5, 48, 62; individualism and, 40; industry and, 37–38, 41, 179; Jacobs and, 101; manufacturing and, 29; Elaine Tyler May on, 30; sociability and, 40; suburbs and, 29–34; urban renewal and, 31–37

Potawatomi Casino, 187

poverty: agriculture and, 117, 125, 150; 1990s and, 57–58, 74, 76, 78; production and, 166, 180; twenty-first century and, 13, 97

Power of Place, The: Urban Landscapes as Public History (Hayden), 60

Pratt, Emmanuel, 142

Presley, Elvis, 204

Prince, Frederick Henry, 168–69

production: activism and, 155, 159, 181, 185; African Americans and, 179, 189; agricul-ture and, 155–61, 166, 169; art and, 227–34; Brooklyn and, 159; Chicago and, 156–61, 165–71; civil rights and, 159, 179; Cleveland and, 155, 158; cre-ative class and, 18, 24, 134, 151, 158, 180, 186, 199, 234, 278; crime and, 192; deindustrialization and, 155, 159, 161, 179, 184, 187; derelict buildings and, 220–35; Detroit and, 155, 158–59, 169–86, 189–92; education and, 165–66, 174, 176; entrepreneurialism and, 158, 171, 178–84, 191–93; Florida and, 158, 180; funding and, 156–59, 161, 167, 180–81, 186, 191–92; government and, 155, 158–59, 184, 189; grants and, 156, 161, 166, 178, 182, 187, 191–92; Great Recession and, 156, 158–60, 168–69, 187; Growing Power and, 135; growth and, 37, 88, 155, 158, 168, 174, 176, 179, 192; housing and, 171, 174, 180, 194; in-dustry and, 155–61, 166–71, 176, 179, 183–84, 187–88; investment and, 155–56, 159, 161, 166, 171, 176, 178, 183; manufacturing and, 158–60, 166–71, 174, 178, 187, 192; markets and, 24, 166, 174, 176, 178; Milwaukee and, 155, 158, 186–89, 192; National Endowment for the Arts (NEA) and, 158, 182, 191; New York and, 155–56, 159, 183; The Plant and, 161–69; police and, 179; post-industrial economy and, 5–6, 15, 18, 53, 57–58, 61, 86, 94, 97, 104, 114–15, 121, 134, 138, 151, 155, 158–78, 187, 198; pov-erty and, 166, 180; Project for Public Spaces (PPS) and, 155, 176; recession and, 159; religion and, 183–92; revital-ization and, 166, 168, 180; rooftop gar-dens and, 140, 155, 198; SHOP and, 239, 242–43; social capital and, 24, 158, 166, 171, 191–92; suburbs and, 187; unemployment and, 178; urban renewal and, 159, 168

Project for Public Spaces (PPS): art and, 4, 197, 211; contemporary placemaking and, 32, 43, 46–47, 51–52; continued relevance of, 90–93; "Downtown Man-agement," 46; Florida and, 87, 90–93, 211; *The Great Neighborhood Book* and, 93; *How to Turn a Place Around*, 91–92; Jacobs and, 91; New York and, 1, 4, 46–47, 70, 87, 93, 155; Nikitin and, 91; Place-making Leadership Council and, 1, 176; production and, 155, 176; "Public Mar-kets as a Vehicle for Social Integration and Upward Mobility," 92; Public Space Assistance Program and, 47; Schum-acher and, 46; twenty-first century and, 86–87, 90–93; Walljasper book and, 93; Whyte and, 2, 32, 43, 46–47, 51–52, 87, 91–92

Project on Regional and Industrial Econom-ics, 100

Project Pop-Up Galleries, 209

Project Row Houses, 199, 211, 257–61, 265, 277

Promise Zones, 102–3

property tax, 113

protests, 14, 21, 78, 117

Pruitt-Igoe, 62

Public Allies Chicago, 18

"Public Markets as a Vehicle for Social Integration and Upward Mobility" report, 92

public space: agriculture and, 123, 155; art and, 198, 200, 213, 221, 238–39; Bryant Park and, 47–48; Central Park and, 52; contemporary placemaking and, 32, 34–35, 40–42, 46–48, 52; developing vi-sions for, 91–92; Jacobs and, 34–35, 40–42, 48–52, 70, 91; land-use planning and, 64; museums and, 35, 62, 187, 204–5, 242, 247, 258; New Urbanism and, 57, 64–68, 76; 1990s and, 59–66, 70, 72, 78; place memory and, 60; police and, 42, 48, 50, 70, 72; power of, 60–62; produc-tion and, 155, 191; Public Space Assis-tance Program and, 47; Schumacher and, 46; size of, 46; sociability and, 40–42; third places and, 3, 48–51, 83; twenty-first century and, 86, 91–93, 104, 106; urban culture and, 60–62; vibrancy and, 103, 113–14, 200; Whyte and, 30–